Human Diversity in North-east India

Human Diversity in North-east India:

Bio-Anthropological Approaches

Edited by

Sarthak Sengupta
Dali Dutta

KALPAZ

First Published, 2021

Published by

Kalpaz Publications
C-30, Satyawati Nagar,
Delhi – 110052
Ph.: 011-47034999, 9811692060
www.kalpazpublications.com
E-mail: kalpaz@hotmail.com

Printed at: Gyan Press, Delhi.

Cataloging in Publication Data—DK
Courtesy: D.K. Agencies (P) Ltd. <docinfo@dkagencies.com>

Human diversity in Northeast India : bio-anthropological approaches / edited by Sarthak Sengupta, Dali Dutta.
pages cm
Contributed articles.
ISBN 9789353246211

1. Biodiversity—Health aspects—India, Northeastern. 2. Physical anthropology—India, Northeastern. 3. India, Northeastern—Scheduled tribes—Health and hygiene. 4. India, Northeastern—Scheduled tribes—Nutrition. I. Sengupta, Sarthak, 1955- editor. II. Dutta, Dali, editor.

LCC QH77.I4H86 2021 | DDC 333.951609541 23

Dedicated
with deep respect to
Prof. R. P. Srivastava and
Prof. K. C. Mahanta

Contents

Preface

North-east India, the erstwhile greater Assam, presently the land of eight sisters states, covers political units namely the States of Arunachal Pradesh, Assam, Manipur, Meghalaya, Mizoram, Nagaland, Sikkim, and Tripura, and is located between latitude 20^0 and 22^0 North and longitude $89^0 46'$ and $97^0 5'$ East. It has an area of 2, 62,179 sq. km., having a hilly and wild topography, and comprises about 7.90 percent landmass of India. The indigenous inhabitants of this region are bewildering in their variety, ethnicity, culture, and folklore. The area is endowed with rich resources of land, water, forests, and minerals, etc., and perched at the confluence of countries like China and Tibet in the north, Myanmar in the east and north, and Bangladesh in the west. This region is connected with the rest of India by a narrow corridor between Nepal and Bangladesh. Because of its geographical features and ethnic composition, North-east India is regarded as a distinct geo-ethnic setting.

The region is a biodiversity hotspot because of its floral and faunal diversity. It is also a cultural hotspot because of its overwhelming cultural and ethnic diversity. There is a rich assembly of cultures and communities in this rugged expanse of the Eastern Himalayas. It is estimated that out of approximately 46 million people living in North-east India. At present, in four of the eight states in North-east India, the tribals are in majority. The average population density of the region is 150 (approx) and the literacy rate of the region is 68.50 percent and the urban population stands at 15.45 percent of the total population.

. The region has been occupied by different streams of the Mongoloid people who came from the north and the east at different periods. The Australoids came to this region before the coming of the Mongoloids who partially or fully absorbed the Australoid strains. At present all the tribes of North-east India are Mongoloid in origin, while the Caucasoid group is formed by the Hindu castes and the Muslims.

In North-east India, the communities living in the valleys are predominantly dependent on agriculture. Rice is the main crop cultivated by the people. Those living in the hill areas of the region mainly practice *jhum* (shifting cultivation). A few small communities living in the extremes of the North-eastern parts of Arunachal Pradesh and the eastern boundary of Nagaland carry on a non-monetized exchange system.

Several languages belonging to different language families are spoken in North-east India. The overwhelming majority of the people are Hindus and constitute the dominant population in Tripura, Assam, and Manipur. The second-largest religious group is the Muslim, which forms a sizable section mainly in Assam. The Christians are mostly concentrated in Nagaland, Mizoram, Meghalaya; and Manipur. Many tribal people of the plains region of Assam and in Arunachal Pradesh have accepted Hinduism. Many others are Hinduised. Although insignificant in number, people having faith in Jainism, Sikhism, and Bahai also exist in the region.

North-east India's diversified culture and people have attracted the attention of many anthropologists from different parts of the globe. This tiny corner of India is a meeting place of so many races and customs that it has long been recognized by anthropologists as one of the most dynamic fields for research work. No area of such limited size can boast of such an impressive array of anthropological findings.

Bio-anthropological studies were introduced in the region at the turn of the twentieth century when Waddell (1901) carried out the first anthropometric works among thirty population groups of Brahmaputra Valley. Since then, most of the bio-anthropological studies are mostly concerned with population variation concerning either morphological, genetical characters, and biochemical markers either singly or in combination without having a conceptual framework and well-formulated research questions to explain the causes of such variation, and it is continuing even today. In North-east India, Mitra (1935, 1936) for the first time reported data on the ABO blood groups for the plains Assamese, Angami, and Lushai populations. Bhattacharjee (1955) initiated dermatoglyphic research in North-east India in the mid-1950s. Population variation in mean age at menarche has been studied first by Rakshit (1960). Das (1966) makes the first move on human growth studies in North-east India. The first treatise entitled '*Ethnic Affinities of the Rabha*' was published by Das (1960). Micro-evolutionary dynamics among several Bodo tribes, Khasi, Caste Assamese, and Koch was studied by Das (1964, 1978, and 1981), Phookan (1975), Sengupta (1982), and many others. Das *et al.,* 1971 proposed a relationship between Austro-Asiatic linguistic affiliation and a high prevalence of HbE among the Khasi of Meghalaya. high prevalence of HbE. Dutta and Kashyap

(2001) studied the population affinities in the region by analyzing three tetrameric STR loci. Mitochondrial DNA and Y chromosome markers among the Khasi and Mishing were analyzed by Reddy *et al.,* (2007), Das and Sengupta (2012).

Despite an early start of anthropological research activities in North-east India, unfortunately till today firm and consequential tradition of research is yet to build up. Whatever has been done on of the people of this easternmost remote corner of the country has not been adequately covered to its vast and numerous unexplored areas. Further, a cursory glance on the repository of bio-anthropological literature from North-east India reveals that information on non-tribal population is too scanty, sporadic, and sketchy to mention its worthiness as well-planned research. Research having social relevance is also conspicuous by its absence. It is often criticized that the anthropological research findings are time and again not practical in approach and therefore cannot be utilized by policymakers. Time has come that the anthropologist having concern about the population of North-east India should focus their attention on making anthropological research more relevant to the survival and well-being of human populations.

It is heartening to note that many physical anthropologists from universities are getting involved in several collaborative efforts with Institutes from mainland India and from abroad. Studies on genetic disorders like thalassaemia, HbS, G-6-P-D also deserve proper attention from the contemporary researcher. It is unfortunate that although several studies have been carried out in North-east India to find out the biological variation between and within the population, there are hardly any studies, which deal with the causes of such variation. The speculation that the micro variation within a given population is due to either mating patterns or geographical isolation is always possible, but it is hardly meaningful without any empirical evidence. Further, in growth and nutritional studies, any attempt to analyze status from somatometric measurements only may not become much meaningful without taking into account evidence from diet analysis, bio-chemical marker, parasitic infestation, etc. Therefore, it warrants more in-depth bio-anthropological studies to understand the causes of variation between and within populations in North-east India. There is also an urgent need for greater integration of cultural and biological anthropology and demands joint efforts to study socially relevant problems. The prospect of biological anthropology will depend upon a thoughtful understanding of its potency and certainly, it lies in the specialized knowledge it can generate.

In this compilation - *Human Diversity in North-east India*, several areas of contemporary biological anthropological research in North-east India have covered: demography (M Sharma; KS Meitei; WAKG Singh and B.Murrey), genetics (SY Meitei and others;), adaptation, human growth, and nutrition (G Begum and P Das; S Banerjee and others; JW Dkhar and N Malsawremkim), somatotype (D Dutta; V Tsukru and JW Dkhar), ageing (C Sharma), reproductive health (J Bora and S Sengupta;) and health and disease (B Das and others; Y Yumnan and HS Singh; DK Limbu and C Lalrammuana; R Deb; K Ghosh and others), metabolic syndrome (G Kamei and G Kshatriya).

We shall consider our efforts to be amply rewarded if this volume in any manner arouses interest and encourage the budding biological anthropologists of India.

Date: 31st December 2020

Sarthak Sengupta
Dali Dutta

Acknowledgements

We have the pleasure to extend our sincere gratitude to all the eminent contributors for their inspiring response who on our request, contributed research articles to this anthology. It is their academic generosity that enabled us to compile this anthology successfully. We have no words to thank them, but we trust this small mention at least records our depth of gratitude for their academic collaboration in making our effort a reality. The names of all contributors are given under the appropriate chapter headings.

We wish to appreciate Dr. Samrat Sengupta, who helped us in designing the jacket. We are also grateful to all those who have helped us directly and indirectly for the successful completion of this treatise.

This book would not have been got prepared in time without the speedy work done by the Director and staff of Kalpaz Publications, New Delhi. We are thankful to all of them.

Date: 31st December 2020

Sarthak Sengupta
Dali Dutta

List of Tables

List of Figures

List of Figures

Contributors

A.K.Gyanendra Singh Wahengbam, Department of Anthropology, University of Delhi, Delhi, 110007.

Abhishikta Ghosh Roy, Anthropological Survey of India, Central Region, Nagpur, India

Anand Sharma, Department of Anthropology, Dibrugarh University, Assam, 786004

Anju Barhai Teli, Department of Biochemistry, Jorhat Medical College, Jorhat, Assam

Arup Ratan Bandyopadhyay, Department of Anthropology, University of Kolkata, 35, Ballygunge Circular Road, Kolkata, West Bengal, 00019

Banrida T. Langstieh, Department of Anthropology, North-Eastern Hill University, Shillong, Meghalaya, 793022

Benrithung Murry, Department of Anthropology, University of Delhi, Delhi, 110007.

Bhaskar Das, Department of Anthropology, DHSK College, Dibrugarh, Assam, 786001

C. Lalrammuana, Department of Anthropology, North-Eastern Hill University, Shillong, Meghalaya, 793022

Chandana Sarmah, Department of Anthropology, Gauhati University, Gopinath Bordoloi Nagar, Jalukbari, Guwahati, Assam, 781014

Dali Dutta, Department of Anthropology, Dibrugarh University, Dibrugarh, Assam, 786004.

Dhruba Kumar Limbu, Department of Anthropology, North-Eastern Hill University, Shillong, Meghalaya, 793022

Diptendu Chatterjee, Department of Anthropology, University of Kolkata, 35, Ballygunge Circular Road, Kolkata, West Bengal, 00019

Gangaina Kamei, Department of Anthropology, University of Delhi, Delhi, 110007.

Gautam Kr. Kshatriya, Department of Anthropology, University of Delhi, Delhi, 110007.

Gulrukh Begum, Department of Anthropology, Gauhati University, Gopinath Bordoloi Nagar, Jalukbari, Guwahati, Assam, 781014

Huidrom Suraj Singh, Department of Anthropology, Manipur University, Chachipur, Imphal, Manipur, 795003.

Johnsly Well Dkhar, Department of Anthropology, North-Eastern Hill University, Shillong, Meghalaya, 793022

Juri Borah, Department of Anthropology, Gurucharan College, Silchar, Assam, 788004.

Kallu Nava Saraswathy, Department of Anthropology, University of Delhi, Delhi, 110007.

Khangembam Somibabu Meitei, Department of Anthropology, Guru Ghasidas Vishwavidyalaya, Bilaspur, Chattisgarh, 495009.

Kusum Ghosh, Department of Anthropology, University of Kolkata, 35, Ballygunge Circular Road, Kolkata, West Bengal, 00019

Maitreyee Sharma, Department of Anthropology, Dibrugarh University, Dibrugarh, Assam, 786004.

Malvika Yumnam, Department of Anthropology, Manipur University, Chachipur, Imphal, Manipur, 795003.

Manoranjan Pal, Economic Research Unit, Indian Statistical Institute, 203 Barrackpore Trunk Road, Kolkata, West Bengal, 700108.

Moajungla Longkumer, Department of Anthropology, North-Eastern Hill University, Shillong, Meghalaya, 793022

Mohinder Pal Sachdeva, Department of Anthropology, University of Delhi, Delhi, 110007.

Nancy Malaswmkim, Department of Anthropology, North-Eastern Hill University, Shillong, Meghalaya, 793022

Nitamoni Bharali, Department of Anthropology, Assam University (Diphu Campus), Karbi Anglong, Assam, 782462.

Nitish Mondal, Department of Anthropology, Assam University (Diphu Campus), Karbi Anglong, Assam, 782462.

Nongthombam Achoubi Devi, Department of Anthropology, Manipur University, Chachipur, Imphal, Manipur, 795003.

Prashanta Kumar Borah, Regional Medical Research Centre (ICMR), Lahoal, Dibrugarh, Assam,

Premananda Bharati, Biological Anthropology Unit, Indian Statistical Institute, 203 Barrackpore Trunk Road, Kolkata, West Bengal, 700108.

Pritam Das, Department of Anthropology, Gauhati University, Gopinath Bordoloi Nagar, Jalukbari, Guwahati, Assam, 781014

Roumi Deb, Amity Institute of Anthropology. Amity University, Noida, Uttar Pradesh

Rumi Deori, Department of Biochemistry, Assam Medical College & Hospital, Dibrugarh, Assam, 786005.

Sanjenbam Yaiphaba Meitei, Department of Anthropology, Manipur University, Chachipur, Imphal, Manipur, 795003.

Sarthak Sengupta, Department of Anthropology, Dibrugarh University, Dibrugarh, Assam, 786004.

Shimul Roy, Department of Anthropology, Vidyasagar University, Medinipur, West Bengal,721102.

Sreeparna Banerjee, Department of Anthropology, West Bengal State University, Berunanpukuria, Malikapur, Barasat, Kolkata, West Bengal, 700126

Subir Biswas, Department of Anthropology, West Bengal State University, Berunanpukuria, Malikapur, Barasat, Kolkata, West Bengal, 700126.

Vivoto Tsukru, Department of Anthropology, North-Eastern Hill University, Shillong, Meghalaya, 793022

Chapter - 1

Nutritional Status of Scheduled Tribe Women of North-east India

Sreeparna Banerjee, Shimul Roy, Manoranjan Pal, Subir Biswas and Premananda Bharati

Introduction

India is home to a large number of tribal populations, many of whom are still unmarked by the modern lifestyle. India has the largest population of tribal people in the world which is about 84.4 million (Baskar and Udhaya, 2018). They constitute 8.6 percent of India's total population (Census of India, 2011). The tribal population of India shows enormous diversity among themselves in respect of their language, ecological settings, physical feature, size of the population, acculturation process, dominant mode of livelihood, social stratification, etc. The majority of scheduled tribe populations are concentrated in the eastern, central, and western belt. Twelve percent of the tribal population are inhabiting in the North-eastern region, five percent in the Southern region, and about three percent in the Northern states (Ministry of Tribal Affairs Government of India, 2014). Schedule tribes are isolated, economically, and socially backward groups. They are excluded from formal education, surrounded by social taboos, poverty, dependency on primitive agriculture (Laxmaiah *et al.,* 2017). Scheduled tribes of India are more susceptible to undernourishment which is recognized as a prevalent health problem. It may be due to the uncertainty of food which results in a serious long-term effect on the health of the tribal populations, especially among children. As a result, it has become a barrier to the development of the nation (Dey and Bisai, 2019).

An impressive study conducted by (Rokade *et al.,* 2020) among tribal women in Maharashtra, finds that over-nourishment has increased significantly with the increase of age group. There have been several research studies describing the nutritional and health status of the tribal population in India (Kshatriya *et al.,* 2016; Verma and Shah, 2014; Das and Bose, 2010; Ghosh and Malik, 2007; Balgir, 2006; Misra, 2004). Malnutrition is one of the major health problems faced by tribal women. Though overweight (12.0%) and obesity (2.0%) are increasing under-weight is still the major concern (37.0% under-weight, among which 8.0 percent severely under-weight; Griffiths and Bentley, 2001). Jaiswal (2018) in his study highlights that Kharwar adults are experiencing serious situations regarding health and nutritional status at all age groups but the women and oldest are suffering most. The nutritional status of scheduled tribe women has been very unsatisfactory for a long period. The nutritional status of young women plays an important role in reproductive performance and work capacity. Lots of improvements have been done regarding the health and nutritional status of the scheduled tribe in North-east states (Prakash, 2016), and a lot more need to be done (Bharali *et al.,* 2017).

Objective

The objective of the present study is to assess the nutritional status of the Scheduled tribe women of the North-east region of India. This study also tries to depict the influence of socio-demographic variables on the nutritional status of the study women.

Methodology

The present study is based on the unit level data extracted from the National Family Health Survey (NFHS-4) conducted by IIPS during 2015-16. This is the largest source that provides women related data at the household level. NFHS-4 provides various socio-demographic information that directly or indirectly influences the nutritional status of women. It is a cross-sectional, comparative as well as ex post facto research. The total sample size consists of 32778 scheduled tribes (ST) women. The area of the present study is the North-eastern region states of India. The study area includes Sikkim, Arunachal Pradesh, Nagaland, Manipur, Mizoram, Tripura, Meghalaya, and Assam. The present study considers only those scheduled tribe Women, who are married and aged between 20 and 49 years. The present study considers only those who have at least one child. Women under 20 years are not included in the present study because for them BMI-for-age is a better indicator than general BMI classification for assessing nutritional status. To study the above-mentioned objectives, appropriate statistical analysis has been used in different steps. Pearson Chi-square test is performed to observe the association between different socio-demographic variables with the nutritional status of women. Binary logistic regressions have been done to see the direct effect of socio-demographic variables on women's nutritional status. Statistical analysis has been carried out using the statistical software package SPSS (18.0) and the significance levels of $p<0.05$ is considered as statistically significant.

Results

Table 1.1 shows the socio-demographic profile of schedule tribe women of North-east India. The highest percentage of women belongs to the 35-39 year group (20.4%), followed by the groups of ages 30-34 years (19.3%), 25-29 years (18.2%), 40-44 years (16.2%), 45-49 years (16.1%) and 20-24 years (9.8%). Educational attainment has been categorized into four groups – no education, primary educated, secondary educated, and highly educated. The majority of women are secondary educated (51.7%). Women with no formal education are 24.0 percent then primarily educated women (19.3%) and only a few women are highly educated (5.1%). More than two-thirds of women reside in rural areas (76.8%). Only 23.3 percent of women reside in urban areas. The majority of women are Christian (71.2%), followed by Hindu (15.4%) and Others (13.4%). 66.2 percent are not working or remain at home. Only 33.8 percent are working. Out of 32,778 women, 43.3 percent of women belong to the poor wealth Index. 27.3 percent of women belong to the middle wealth index. More than one-fourth of women (29.4%) belong to the higher wealth index category.

Table 1.1: Socio-demographic profile of Scheduled Tribe women of North-eastern India

Socio-demographic variable	***Percentages***
Age	
20-24	3197 (9.8)
25-29	5966 (18.2)
30-34	6322(19.3)
35-39	6700 (20.4)
40-44	5302(16.2)
45-49	5291 (16.1)
Educational attainments	
No education	7862 (24.0)
Primary	6310 (19.3)
Secondary	16948 (51.7)
Higher	1658 (5.1)

(Cond...)

Residence	
Urban	7613 (23.2)
Rural	25165 (76.8)
Religion	
Christian	23338 (71.2)
Hindu	5036 (15.4)
Others	4406 (13.4)
Working Status	
Not working	3629 (66.2)
Working	1849 (33.8)
Wealth Index	
Poor	14191 (43.3)
Middle	8958 (27.3)
Rich	9629 (29.4)

Figures within parentheses indicate percentages

Table 1.2 depicts that younger women aged 20-24 years have a higher percentage of underweight (13.7%) compared to women aged 25-29 years (10.3%) and above age groups. As age increases, the percentage of underweight decreases, and the percentage of overweight and obesity increases. A higher percentage of overweight and obesity was found among women aged above 30 years. Chi-square test shows that these differences are statistically significant at a 1.0 percent level. The table also points out that women with primary education and no formal education are more like to be underweight (9.9% and 9.5%) than women who are secondary (7.8%) and highly educated (7.2%). Overweight and obesity are predominant among women who have higher education (31.6%). Chi-square test reveals that women with better educational attainment have better nutritional status and the result is statistically significant. A higher percentage of underweight can be observed among women who belong to the Hindu religion (11.1%), which is followed by Christian (8.5%) and other religions (6.1%). A higher prevalence of overweight and obesity was observed among women who belong to 'Other' religions (28.0%) which is followed by Christian (19.6%) religion. Chi-square test shows religion has a statistically significant influence on the nutritional status of women.

Women who are not working or remain at home are more prone to be underweight (8.6%) than women who are working (7.8%). Overweight and obesity among working and not working women are more or less the same, though the working women show a higher percentage of overweight and obesity (21.8%) than not working women (21.0%). Chi-square test shows no significant association between nutritional status and working status of women. The wealth index shows a significant association with

Table 1.2 Association of socio-demographic indicators and nutritional status of study women

Socio-demographic Variable	*BMI*			χ^2
	Underweight (%)	*Normal (%)*	*Overweight/ obesity*	
Age groups				
20-24	431 (13.7)	2419 (76.9)	295 (9.4)	718.114**
25-29	600 (10.3)	4430 (75.7)	820 (14.0)	
30-34	467 (7.5)	4465 (72.0)	1273 (20.5)	
35-39	490 (7.5)	4582 (69.8)	1493 (22.7)	
40-44	382 (7.4)	3447 (66.4)	1364 (26.3)	
45-49	385 (7.4)	3429 (66.1)	1373 (26.5)	
Total	2755 (8.6)	22772 (70.8)	6621 (20.6)	
Educational attainment				
No education	733 (9.5)	5684 (73.9)	1275 (16.6)	331.512**
Primary	615 (9.9)	4577 (73.7)	1022(16.4)	
Secondary	1290 (7.8)	11518 (69.3)	3808 (22.9)	
Higher	117 (7.2)	993 (61.2)	512 (31.6)	
Total	2755 (8.6)	22772(70.8)	6617 (20.6)	
Residence				
Urban	531 (7.1)	4696 (62.9)	2237 (30.0)	526.270**
Rural	2224 (9.0)	18076 (73.2)	4380 (17.7)	
Total	2755 (8.6)	22772 (70.8)	6617 (20.6)	
Religion				
Christian	1942 (8.5)	16445 (71.9)	4483 (19.6)	222.429**
Hindu	550 (11.1)	3468 (70.2)	922 (18.7)	
Others	263 (6.1)	2859 (66.0)	1212 (28.0)	
Total	2755 (8.6)	22772 (70.8)	6617 (20.6)	
Working status				
Not working	307 (8.6)	2507 (70.4)	748 (21.0)	1.372
Working	141 (7.8)	1277 (70.4)	395 (21.8)	
Total	448 (8.3)	3784 (70.4)	1143 (21.3)	
Wealth Index				
Poor	1479 (10.6)	10769 (77.4)	1666 (12.0)	1596.563**
Middle	728 (8.3)	6240 (71.0)	1817 (20.7)	
Rich	548 (5.8)	5763 (61.0)	3134 (33.2)	
Total	2755 (8.6)	22772 (70.8)	6617 (20.6)	

Figures within parentheses indicate percentages ; P<0.01**, p<0.05 *

nutritional status. Women with poor wealth index show a higher percentage of underweight (10.6%), which is followed by the middle wealth index (8.3%). Underweight is the lowest among the richer wealth index category. The result implies that the economic condition of women is very important for better nutrition of women. In contrast to the result of underweight, overweight, and obesity (33.2%) are highest among richer wealth-index category women.

Binary logistic regressions have been carried out to see whether some categories of women (age, educational attainments, working status, type of earnings, and wealth-index categories) have a higher vulnerability of undernourishment than others. Socio-demographic variables are used as independent variables and the nutritional status of women is considered as a dependent variable. Binary code is used in the dependent variable, where underweight is coded as '0' and Normal (others, who are not underweight) were coded as '1'.

Table 1.3: Estimate effects and a significance level of selected socio-demographic variables on the nutritional status of North-eastern scheduled tribe women, aged 15-49 years

Socio-demographic Variables	*Exp (B)*	*95% C.I. for EXP (B)*	
		Lower	*Upper*
Age			
20-29®			
30-39	1.613**	1.276	2.039
40-49	1.613**	1.248	2.083
Educational attainment			
Not education ®			
Primary	1.061	0.799	1.408
Secondary	1.344*	1.036	1.744
Higher	1.150	0.659	1.164
Residence			
Rural ®			
Urban	0.870	0.651	1.164
Religion			
Christian®			
Hindu	0.791	0.620	1.011
Others	1.839**	1.289	2.624
Working status			
Not working ®			
Working	1.050	0.848	1.301
Wealth Index			
Poor ®			
Middle	0.979	0.773	1.240
Rich	1.709**	1.237	2.360
Constant	6.494		
*p<0.05; **P<0.001			

Table 1.3 indicates that women aged 30-39 years and 40-49 years are more likely (1.613 times) to be not underweight than 20-29 years women and the result is statistically significant. Women with a secondary level of education are 1.344 times more likely to be normal compared to women with no formal education. Hindu women are more likely to become underweight than Christian women. Women belonging to other religious groups are more likely (1.839 times) to have non-underweight compared to Christian women and it is highly significant. Working women tend to become underweight (0.828 times) than not working women. Wealth Index

shows a statistically significant association with the nutritional status of women. It shows that women belonging to richer wealth are 1.709 times more likely to be non-underweight than poor wealth index category women.

Discussion

The prevalence of underweight among women is considered a serious public health problem in developing countries like India. In India, the majority of the population is malnourished and underprivileged (Mondal and Sen, 2010; Nandy *et al.,* 2005). In the present study, a significant association is found between women's age group and the nutritional status of women. Similar results were found in the study of Nayak *et al.,* (2016). Bose and Chakraborty (2005) depict that underweight among adult women of Bathudis is high. Philomenamma and Ramdas (2008) observe more undernutrition among women (52.2%) than men (23.1%). The study of Barbhuiya and Das (2013) among Meiteis of Cachar district of Assam finds that a high prevalence of undernutrition exists among adults (more than 30.0%). Bisai *et al.,* (2008) investigate nutritional status among Savar women and point out that more than 50.0 percent of women belong to the underweight category. The nutritional status of women in the present study is better compared to other studies. In the present study, only 8.4 percent of adult women aged 20-49 years are underweight.

Table 1.4: Comparative account of mean BMI of Scheduled Tribal women in different studies

State	*Community*	*BMI (Mean)*	*Reference*
West Bengal	Bhumij	18.4	Ghosh (2007)
West Bengal	Oraon	19.7	Mittal and Sivastava (2006)
Odisha	Juang	18.3	Goswami (2013)
	Bathudis	17.9	Bose and Chakraborty (2005)
Assam	Karbis	21.53	Goswami and Bhattacharyya (2015)
Uttar Pradesh	Kharwar	20.5	Jaiswal (2018)
Andhra Pradesh	Bhagat	22.21	Vardrajan and Prasad (2009)
	Konda Dora	21.55	
	Gonds	22.38	
	Raj Gonds	22.51	
	Sugali	23.27	
	Yerrukula	23.68	
North-eastern states		22.60	Present study

Vardarajan and Prasad (2009), in their study among Andhra Pradesh tribal women, find the BMI of women ranging from 21.55 to 23.68, which are similar to the present study. In the present study, the mean BMI is 22.60 among North-eastern tribal women. Their mean BMI is comparatively higher than most other tribal groups. Bhathudis (17.9), Juang (18.3), Bhumij (18.4), and Oraon (19.7) have the lowest mean BMI (less than 20.0 Kg/m^2) as shown in table 1.4. Poluru and Jareena (2009) depict that women aged 25-34 years and 35-49 years in Gujarat and Maharashtra have more tendency to became overweight or obese than women in the youngest age group of 15 to 19 years. In the present study also, it is found that younger women are more likely to become underweight. As age increases women are more likely to become overweight and obese. The association and effect of age with nutritional status are statistically significant according to chi-square and regressions. Mungreiphy and Kapoor (2010) studied Tangkhul Naga women in Manipur and report that 25.10 percent of tribal women were overweight and 2 percent were obese. Sen *et al.*, (2013) mentioned a higher prevalence of overweight and obesity (20.33 percent and 29.33 percent respectively) among the Bengalee ethnic group of West Bengal. Subramanian *et al.*, (2007) in their study points out that 'overweight and obesity' was higher among the richer wealth index category compared to the poorer sections. Griffiths and Bentley's (2001) study on women in Andhra Pradesh and find that 12.0 percent of women are overweight and 2.0 percent were obese. In the present study also, we have found that richer wealth index women have a higher percentage of overnutrition and the wealth index has a significant association with the nutritional status of women.

Table 1.5: Comparative account of overweight and obesity among tribal women in different studies

State	*Community*	*Overnutrition*		*Reference*
		Overweight (%)	*Obesity (%)*	
Arunachal Pradesh	Nyishi	9.94	9.57	Bharali *et al.*, 2017
Assam	Karbi	17.33	14.33	Mondal *et al.*, 2017
Assam	Regma Naga	25.50	11.63	Rengma *et al.*, 2015
West Bengal	Bhuijs	17.7		Acharaya 2016
West Bengal	Dhodias	23.8		
West Bengal	Kuknas	15.8		
	Indian tribes	10.9	1.5	
Manipur	Tangkhul Naga	25.10	2	Mungreiphy and Kapoor 2010
Present study		20.1	0.4	

Table 1.5 shows the percentage of overweight and obesity among tribal women in different studies. The highest percentage of overweight are found to be among Rengma Naga (25.50% by Rengma *et al.*, 2015), which is

followed by Tangkhul Naga women (25.10%; Mungreiphy and Kapoor 2010). Overweight and obesity among Nyishi tribal women are 9.94 percent and 9.57 percent respectively. Mondal *et al.,* (2017) in their study among Karbi women of Assam have pointed out that 17.33 percent and 14.33 percent of women are overweight and obese. An impressive study among Dhodias, Bhuijs, Kuknas, and Indian tribes women has been conducted by Acharaya (2016), to find the percentages of overweight women as 23.8 percent, 17.7 percent, 15.8 percent, and 10.9 percent respectively, which are very alarming. The present study also depicts high percentages of over nourishment in North-east Indian tribal women. The prevalence of overweight and obesity among North-eastern tribal women was 20.1 percent and 0.4 percent respectively in the present study.

Table 1.6: Comparative account of underweight among tribal women in different studies

State	*Community/ Area*	*Underweight*	*Reference*
Jharkhand	Orang	62.50	Dutta, 2011
India	Indian Tribes	47.40	Kshatriya *et al.,* 2016
Arunachal Pradesh	Nyshi	10.50	Bharali *et al.,* 2017
West Bengal (Kolkata City)	Munda	58.5	Ghosh and Bharati 2006
West Bengal	Santal	46.7	Das and Bose 2010
	Kora Mudi	52.2	Bose *et al.,* 2006
Odisha	Bathudi	64.50	Bose and Chakraborty 2005
West Bengal	Dhimal	36.7	Banik *et al.,* 2007
Present study		8.6	

Table 1.6 shows the undernourishment among tribal women in different studies. In this table, the highest percentage of underweight can be observed among Bathudi tribe (64.50%; Bose and Chakraborty, 2005) then followed by Orang (62.50%; Dutta 2011), Munda (58.5%; Ghosh and Bharati 2006), Kora Mudi (52.2%; Bose *et al.,* 2006), Indian tribe (47.40%; Kshatriya *et al.,* 2016), Santal (46.7%; Das and Bose 2010), Dhimal (36.7%; Banik *et al.,* 2007), Nyshi (10.50%; Bharali *et al.,* 2017) and in the present study (8.6%). The table also depicts that in the present study, the North-eastern tribal women have a lower percentage of underweight than other studies. So, overweight and obesity are more serious among the study population compared to undernourishment.

Conclusion

The nutritional status of any community or population is very important as it represents the overall health status. As women bear the future generation, a woman with normal nutritional status will increase the chance to give birth to a healthy baby. But in the present study, young age women aged below 30 years are found to be more prone to be underweight. As age increases, women are more likely to be overweight and obese. If the younger age women become the victim of underweight, then their future generation is likely to have poor nutritional and health status. In the present study, the percentage of underweight is lower than most other studies conducted among tribal women in other parts of India. But at the same time, the percentage of overweight and obese is higher which is alarming. Overweight and obesity situation among women is more serious in the present study than the situation of underweight. Overweight and obese come with many health complications. As many other studies suggest that an increasing trend of overweight and obesity, if not checked, may become very problematic soon. Other studies need to be conducted to understand the situation in depth.

Acknowledgements

The authors are grateful to DHS International who permitted and provided us the data of the Fourth National Family Health Survey, 2015-16. Without this data support, the present study could not be possible to carry out.

References

Acharaya, SK 2016, 'The tribal burden of obesity, undernutrition, and cardiovascular disease risk among Indian tribes', *PloS One*, vol.11, no.1, pp.1-19.

Balgir, RS 2006, 'Tribal health Problem, disease burden, and ameliorative challenges in tribal communities with special emphasis on tribes of Odisha', *Proceeding of National Symposium on Tribal Health*, pp.161-176. Accessed on 3.6.2020. https://www.nirth.res.in/publications/nsth/22.RS.Balgir.pdf

Banik, D, Bose, K, Bisai, S, Bhattacharya, M & Das, S 2007, 'Chronic energy deficiency among adult Dhimals of Naxalbari, West Bengal: Comparison with other tribals of Eastern India', *Food and Nutrition Bulletin*, vol.28, no.3, pp.348-352.

Barbhujya, AFG & Das, R 2013, 'Gender differences in nutritional status among the adult Meiteis of Cachar district of Assam, India', *Eurasia Journal of Anthropology*, vol. 4, no. 2, pp. 36-44.

Baskar, R & Udhaya, R 2018, 'A study on health status of Irula tribes in Mullankadu Coimbatore district, Tamilnadu, India', *International Journal of Social Science and Economic Research,* vol.12, no.3, pp. 7192-7198.

Bharali, N, Mondal, N & Singh, N 2017, 'Prevalence of undernutrition, overweight, and obesity among Nyishi tribal women of Arunachal Pradesh, North-east India', *Human Biology Review*, vol.6, no.1, pp. 63-78.

Bisai, S & Bose, K 2008, 'Body mass index and chronic energy deficiency among adult tribal population of West Bengal: A review', *Tribes and Tribals*, Special. Vol.2, pp. 87-94.

Bose, K & Chakrabarty, F 2005, 'Anthropometric characteristics and nutritional status based on body mass index of adult Bathudis: A tribal population of Keonjhar district, Odisha, India', *Asia Pacific Journal of Public Health*, vol.14, no.1, pp.80-82.

Bose, R, Ganguli, S, Mamtaz, H, Mukhopadhyay, A & Bhadra, M 2006, 'High prevalence of undernutrition among adult Kora Mudi tribals of Bankura District, West Bengal, India', *Anthropological Science,* vol.114, no.1, pp. 65-68.

Census of India 2011, *Office of the Registrar General and Census Operation*, Ministry of Home Affairs, Government of India, New Delhi, India.

Das, S & Bose, K 2010, 'Body mass index and chronic energy deficiency among adult Santals of Purulia District, West Bengal, India', *International Journal of Human Science*, vol.7, no.2, pp.488-503.

Datta, B 2011, 'Nutritional status adiposity and body composition of Oraon and Sarak females in Ranchi District, India- a comparison', *Ecology of food and nutrition,* vol.50, no.1, pp. 43-62.

Dey, U & Bisai, S 2019, 'The prevalence of undernutrition among the tribal children in India: a systematic review', *Anthropology Review*, vol.82, no.2, pp.203-217.

Ghosh, M 2007, 'Nutritional status of adult Bhumij males of Kharagpur, Paschim Medinipur,' Paper presented in a seminar on *Approaching Development in Department of Anthropology.* Vidyasagar University, India. 5,1:22.

Ghosh, R & Bharati, P 2006, 'Nutritional status of adults among Munda and Pod populations in a peri-urban area of Kolkata City, India,' Asia Pacific Journal of Public Health, vol.18, no.2, pp.12-20.

Ghosh, S & Mallik, SL 2007, 'Sex differences in body size and shape among Santhal of West Bengal,' *Anthropologist*, vol.9, no.2, pp. 143-149.

Goswami, M 2013, 'Prevalence of under-nutrition among the Juang Physical Anthropology', *Antrocom*, vol.9, no.1, pp. 61–66.

Goswami, R & Bhattacharya, M 2015, 'Nutritional status among adult Karbi women of Kamrup District, Assam', *International Research Journal of Social Sciences*, vol.4, no.9, pp.17-20.

Griffiths, PL & Bentley, ME 2001, 'The nutrition transition is underway in India' *Nutrition*, vol.131, no.10, pp. 2692–2700.

Jaiswal, A. 2018, 'Nutritional and health status evaluation of tribes of Uttar Pradesh: An anthropological dimensions', *Global Journal of Archaeology and Anthropology,* vol.2, pp.0042-0047.

Kshatriya, GK & Acharya, SK 2016, 'Gender disparities in the prevalence of undernutrition and the higher risk among the young women of Indian tribes, *PLoS One*, vol.11, no.7, e0158308.

Laxmaiah, K, Rao, M, Kumar, H, Arlappa, N, Venkaiah, K & Brahmam, GNV 2017, 'Diet and nutritional status of tribal Population in ITDA project areas of Khammam District, Andhra Pradesh', *Journal of Human Ecology*, vol.21, no.2, pp.79-86.

Ministry of Tribal Affairs Government of India 2014, Report of the high-level committee on socio-economic health and educational status of tribal communities of India.

Misra, P 2004, *Ecology, Culture, and Health Primitive Tribe*. New Delhi: Serials Publications.

Mittal, PC & Srivastava, S 2006, 'Diet, nutritional status, and food-related traditions of Oraon tribes of New Mal (West Bengal), India', *Rural and Remote Health*, vol.6, no.1, p.385.

Mondal, N & Sen, J 2010, 'Prevalence of under-nutrition among children (5–12 years) belonging to three communities residing in a similar habitat in North Bengal, India', *Annals of Human Biology Review*, vol.37, no.2, pp.199–217.

Mondal, N, Timungpi, R, Sen, J & Bose, K 2017, 'Socio-economic and demographic correlates of overweight and obesity: A study on the Karbi women of Assam North-east India', *Anthropology*, vol.2, no.2, pp. 31-39.

Mungreiphy, NK & Kapoor, S 2010, 'Socio-economic changes as covariates of overweight and obesity among Tangkhul Naga tribal women of Manipur, North-east India', *Journal of Bio-social Science*, vol.42, no.3, pp. 289-305.

Nandy, S, Irving, M, Gordon, D, Subramanian, SV & Smith, GD 2005, 'Poverty, child undernutrition, and morbidity: new evidence from India', *Bull World Health Organisation,* vol.83, no.3, pp.210–216.

Nayak, M & Sreegiri, S 2016, 'A study on nutritional status of tribal women in Visakhapatnam district, Andhra Pradesh, India', *International Journal of Community Medicine and Public Health,* vol.3, no.8, pp.2049–2053.

Philomenamma, J & Ramadas, S 2008, 'Body mass index: an indicator of nutritional status among adult Mannan tribes of Idukki district, Kerala', *SB academic review*, vol. XV, pp.60–65.

Poluru, R & Jareena, S 2009, 'Overweight exceeds underweight among women in Kerala: An analysis of trends and determinants,' *Journal of Human Ecology*, vol.25, no.2, pp.93-103.

Prakash, A. 2016, 'Health in North-east region of India- the new focus of attention', *Indian Journal of Medical Specialities*, vol.7,(2016), pp.93-94.

Rengma, MS, Sen, J & Mondal, N 2015, 'Socio-economic, demographic and lifestyle determinants of overweight and obesity among adults of North-east India', *Ethiopian Journal of Health Sciences*, vol.25, no.3, pp.199-208.

Rokade, S, Mog, M & Mondal, NA 2020, 'Nutritional status among tribal women in Maharashtra, India: Spatial variations and determinants', *Clinical Epidemiology and Global Health*, 2213-3984. In press

Sen, J, Mondal, N & Dutta, S 2013, 'Factors affecting overweight and obesity among urban adults: a cross-sectional study', *Epidemiology and Biostatistics Public Health,* 10(1): e8741-1 –e8741-11.

Subramanian, SV, Kawachi, I & Smith, GD 2007, 'Income inequality and the double burden of under- and overnutrition in India', *Journal of Epidemiology and Community Health,* vol.61, no.9, pp.802-809.

Varadarajan, A & Prasad, S 2009, 'Regional variations in nutritional status among tribals of Andhra Pradesh', *Studies of Tribes and Tribals*, vol.7, no.2, pp.137-141.

Verma, M & Shah, A. 2014, 'Health, tradition, and awareness: A perspective on the tribal health care practices', *Research Process*, vol.2, no.2, pp.82-91.

Chapter - 2

Patterns of Growth and Maturity Gradient among the Aka Adolescent Boys of Arunachal Pradesh

Gulrukh Begum and Pritam Das

Introduction

There are nearly 1.2 billion adolescents (10-19 years old) worldwide. In some countries, adolescents make up as much as a quarter of the population and the number of adolescents is expected to rise through 2050, particularly in low- and middle-income countries (LMICs) where close to 90.0 percent of 10- to 19-years-olds live. This section of the population therefore, currently constitutes a demographic bulge. This phase of life is also described as 'demographically dense', a period in life during which a large percentage of people experience a large percentage of key life-course events (Rindfuss, 1991).

This is the period of transition between childhood and adulthood - a time of rapid physical, cognitive, social, and emotional maturation as the boy prepares for manhood. The precise boundaries are difficult to define. It has three sub-phases: early adolescence, middle adolescence, and late adolescence. The changes that occur during adolescence as a part of the pubertal process includes maturational and hormonal. This is also the population group that remains overlooked and neglected in nearly every health care system. An estimated 1.2 million adolescents die each year, largely from preventable causes.

Growth during adolescence is faster than at any other time in an individual's life except for the first year. Good nutrition during adolescence is critical to cover the deficits suffered during childhood and should include nutrients required to meet the demands of physical and cognitive growth and development, provide adequate stores of energy for illnesses and pregnancy, and prevent adult-onset of nutrition-related diseases (WHO, 2006). Studies conducted in different countries in the South Asian region reveal that nutritional deprivation affects almost all growth parameters and final adult body size resulting in thinness and stunting.

WHO (2006) has therefore, stated that closing the gaps, both in research and in action, would benefit society as a whole, resulting in improved health and nutrition of adolescents and help in harnessing their full physical and mental potential for the overall improvement of the populations and economies. In the present study, it is therefore, intended to study the growth pattern of the Aka boys of Arunachal Pradesh from 9 to 18 years of age. The bodily proportions of a child change with the advancing age. An estimation of this change provides a measure of maturity. The present study also focuses on the maturity gradient of all the linear, circumferential, and weight measurements under consideration.

Materials and method

People

Aka is one of the scheduled tribes of Arunachal Pradesh. Of the 26 major tribes and more than 100 sub-tribes in the state, Akas are one of the lesser-known tribes in the entire North-east. The Aka, also known as *Hrusso*, are found in the Thrizino (cultural hub), Bhalukpong (commercial hub), Buragaon, Jamiri, Palizi, Khuppi area in West Kameng of the Indian state of Arunachal Pradesh. They speak a language that belongs to the Tibeto-Burman family. The Aka share strong cultural affinities with the Miji tribe, and intermarriage with the Miji is also prevalent. As per the census of India 2011, the total no. of Aka people in Arunachal Pradesh is 8167.

The Akas are divided into two major groups, *Hrusso Aka* and *Koro Aka* (also known as *Bana Aka*). The *Koro Akas* are generally settled in the East Kameng district and *Hrusso Akas* are settled in the West Kameng district. Some of the Aka clans are *Aglasow, Ramdasow, Sakrinsow, Richisow, Sagro, Delusow, Sidisow, Nimasow, Parisow,* etc. The data for the present study was collected from the Thrizino circle of West Kameng district. Thrizino is located at a latitude of 28°21' N and a longitude of 94°72' E. Thrizino township is the most developed area of the entire circle and it is surrounded by villages and *bastis*. Many people have their houses in Thrizino Township as well as in the nearby *basti*. Data was collected from Thrizino Township as well as three nearby villages, viz: Giziri, Subu, and Palizi respectively.

Data collection was carried out from 26th March to 8th April 2018. A sample of 300 boys has been measured for eight different body dimensions. The whole sample of 300 boys is further divided into ten age groups from 9 to 18 years at an interval of one year. Special care was taken so that each age group had a minimum of 30 children.

Anthropometric measurements

Stature, sitting height, lower extremity length, head circumference, bicep girth, chest girth, calf girth was measured in centimeters with the help of an anthropometer and cloth tape whereas body weight was measured in kilograms with the help of a weighing scale. The standard method of Weiner and Lourie (1981) was followed for taking the measurements. The height of the children was recorded to the nearest 0.10 cm. The weight was measured with bare feet of the same children wearing minimum clothing. Weight was recorded to the nearest 0.50 kg.

The age of the subjects was ascertained from birth certificates or official school records. The collected sample comes under the 95.0 percent confidence interval of the total Aka population of the region. Technical error of measurement was also calculated which was found to be within the limit.

Results

The descriptive statistics for linear, circumferential, and weight measurements of the Aka boys are given from table: 2.1 to 2.8 and figure: 2.1 to 2.8.

The mean values of all the body dimensions understudy show an increase with the advancing age. From the distance curve (Figure: 2.1) as well as from the descriptive statistics (Table:2.1) of stature, it could be seen that the maximum gain in stature is between 13 to 14 years (8.41cm). The overall gain over ten years is 39.21 cm or 27.69 percent. The values of maturity gradient show that the boys attain 93.56 percent growth in stature at the age of 14 years. At 15 years the amount of growth completed in stature is 98.06 cm and after 15 years, only 1.94 cm of growth in stature remains to be completed. Similar is the case with the other body measurements under study. In sitting height, the highest gain is between 13 and 14 years whereas it is between 14 and 15 years in lower extremity length. The total gain in ten years from 9 to 18 years is 28.11 cms in sitting height and in lower extremity length it is 11.1 cms. So, the sitting height grows more than the lower extremities during the adolescent period. At 16 years in both sitting height and lower extremity length, more than

98.0 percent of growth is already complete among the Aka boys. Only 1.87 percent of growth remains to be completed in sitting height and it is only 1.64 percent in lower extremity length. At 9 years the sitting height has grown only 64.73 percent and the lower extremity has already grown 85.94 percent.

Table 2.1: Descriptive statistics of the stature of the Aka boys of Arunachal Pradesh

Age in years	*n*	*Mean ±S.E.*	*S.D ±S.E.*	*C.V.±S.E.*	*Growth*		*Growth gradient*
					Absolute	*% per annum*	
9+	32	119.50±0.72	4.00±0.51	3.42±0.42			75.29
10+	33	123.85±1.01	5.83±0.71	4.71±0.58	4.35	3.64	78.03
11+	31	129.31±1.04	5.84±0.74	4.52±0.57	5.46	4.22	81.47
12+	31	134.72±1.03	5.78±0.73	4.29±0.54	5.41	4.01	84.88
13+	27	140.08±1.78	9.28±1.26	6.63±0.90	5.36	3.82	88.26
14+	28	148.49±1.40	7.42±0.99	4.99±0.66	8.41	5.48	93.56
15+	26	155.64±1.58	8.09±1.12	5.20±0.72	7.15	4.59	98.06
16+	29	155.93±1.27	6.89±0.90	4.42±0.58	0.29	0.18	98.24
17+	30	158.05±1.02	5.62±0.72	3.55±0.45	2.12	1.34	99.58
18+	33	158.71±1.07	6.16±0.75	6.16±0.47	0.66	0.41	100

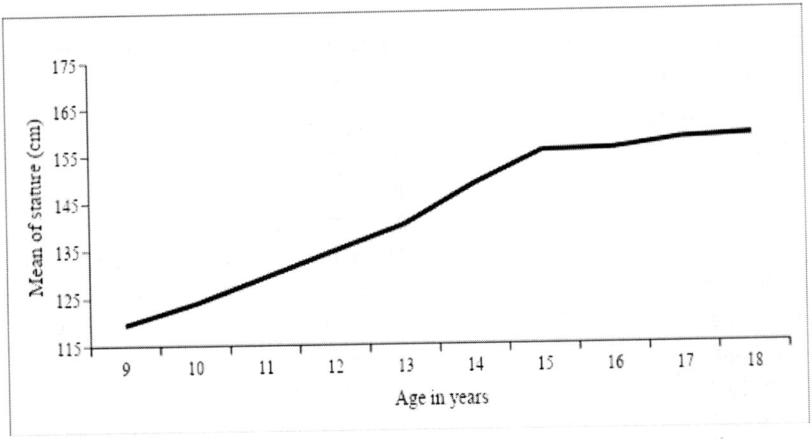

Figure 2.1: Distance curve of the stature among the Aka boys of Arunachal Pradesh

Table: 2.2. Descriptive statistics of the sitting height of the Aka boys of Arunachal Pradesh

Age in years	*n*	*Mean ± S.E.*	*S.D.±S.E.*	*C.V±S.E.*	*Growth*		*Growth gradient*
					Absolute	*% per annum*	
9+	32	51.60±0.71	4.06±0.50	7.87± 0.98			64.73
10+	33	54.86±0.88	5.10±0.62	9.30±1.14	3.26	6.31	68.82
11+	31	60.13±1.14	6.20±0.78	10.32±1.13	5.27	9.60	75.43
12+	31	63.85±0.96	5.37±0.68	8.41±1.06	3.72	6.18	80.10
13+	27	68.82±1.62	8.41±1.14	12.23±1.66	4.97	7.78	86.33
14+	28	74.30±1.12	5.94±0.47	8.00±1.06	5.48	7.96	93.21
15+	26	78.00±1.26	6.43±0.89	8.24±1.14	3.70	4.97	97.85
16+	29	78.22±1.17	6.32±0.83	8.00±1.06	0.22	0.28	98.13
17+	30	79.30±0.79	4.34±0.56	5.48±0.70	1.08	1.38	99.98
18+	33	79.71±0.71	4.11±0.50	5.16±0.63	0.41	0.51	100

Table 2.3: Descriptive statistics of lower extremity length of the Aka boys of Arunachal Pradesh

Age in years	*n*	*Mean±S.E*	*S.D ±S.E*	*C.V±S.E*	*Growth*		*Growth gradient*
					Absolute	*% per annum*	
9+	32	67.9±0.55	3.11±0.38	4.57±0.57			85.94
10+	33	68.99±0.49	2.81±0.34	4.07±0.50	1.09	1.60	87.32
11+	31	69.72±0.52	2.90±0.36	4.19±0.53	0.73	1.05	88.25
12+	31	70.87±0.72	4.02±0.51	5.68±0.72	1.15	1.64	89.70
13+	27	71.26±0.67	3.50±0.47	4.92±0.66	0.39	0.55	90.16
14+	28	74.19±0.67	3.58±0.47	4.83±0.64	2.93	4.1[illegible]	93.91
15+	26	77.64±0.87	4.46±0.61	5.75±0.79	3.45	4.65	98.27
16+	29	77.71±0.71	3.86±0.50	4.97±0.65	0.07	0.09	98.36
17+	30	78.75±0.73	4.00±0.51	5.09±0.65	1.04	1.33	99.68
18+	33	79.00±0.89	5.14±0.63	6.51±0.80	0.25	0.31	100

The percentage of adult value attained by the Aka boys for head circumference at 9 years is 92.75 percent. So, at 9 years more than 90.0 percent of the growth in head circumference is already over (Table: 2.4). Only 7.25 percent of growth remains to be completed. The maximum increase takes place between 13 and 14 years which is 0.78 cms. The overall gain during the period of 10 years from 9 years to 18 years is only 4 cms. This is because the maximum amount of growth in head circumference is already over at the birth of an individual.

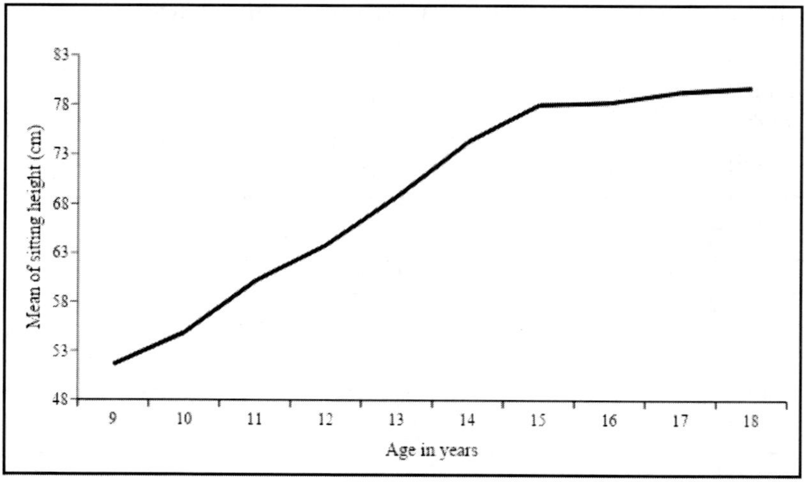

Figure: 2.2 Distance curve of sitting height among the Aka boys of Arunachal Pradesh

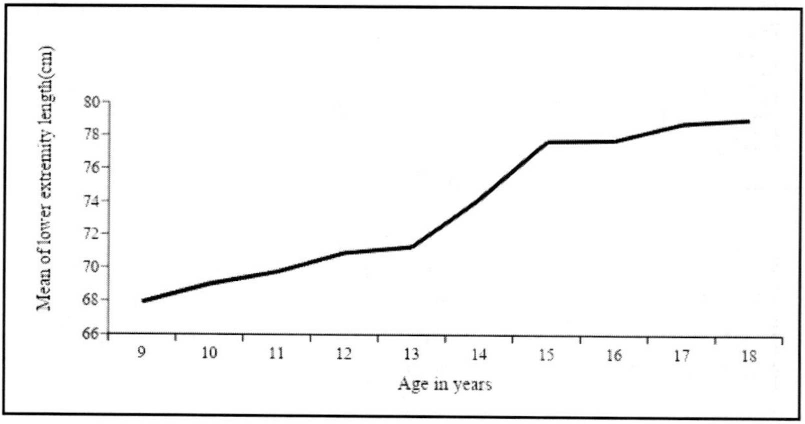

Figure: 2.3 Distance curve of lower extremity length among the Aka boys of Arunachal Pradesh

The maturity gradient for the measurements under consideration of the Aka boys at each age is presented in table: 2.1 to 2.8. At 15 years, in all the measurements under study 90.0 percent of growth is already over, i.e. in all the dimensions, 90.0 percent of maturity level is already reached. Of all the measurements under study, it is in the circumferential measurements where maximum growth remains to be completed at 9 years. In girth of biceps, it is 32.67 percent in chest circumference it is 26.36 percent and in calf circumference, it is 27.26 percent of growth remains to attain maturity status. Bodyweight is a measurement that reflects any change in any part of the body either increases or decreases. The descriptive statistics of body weight (Table: 2.8) and the distance curve (Figure: 2.8) show an increase with the advancing age. From the distance curve and Table: 2.8, it could be seen that the maximum gain is between 14 and 15 years (6.87 kg). The overall gain over ten years is 29.59 kg or 56.47 percent maturity status.

Table 2.4: Descriptive statistics of the head circumference of the Aka boys of Arunachal Pradesh

Age in years	*n*	*Mean± SE*	*SD ±SE*	*CV±SE*	*Growth*		*Growth gradient*
					Absolute	*% per annum*	
9+	32	51.20±0.23	1.32±0.16	2.58±0.32			92.75
10+	33	51.56±0.28	1.65±0.20	3.20±0.39	0.36	0.7	93.40
11+	31	51.65±0.31	1.74±0.22	3.37±0.42	0.09	0.17	93.56
12+	31	52.11±0.26	1.44±0.18	2.77±0.35	0.46	0.89	94.40
13+	27	52.75±0.39	2.06±0.28	3.91±0.53	0.64	1.22	95.56
14+	28	53.53±0.26	1.41±0.18	2.63±0.35	0.78	1.47	96.97
15+	26	54.27±0.27	1.41±0.19	2.60±0.36	0.74	1.38	98.31
16+	29	54.57±0.24	1.32±0.17	2.42±0.31	0.30	0.55	98.85
17+	30	54.66±0.22	1.23±0.15	2.26±0.29	0.09	0.16	99.02
18+	33	55.20±0.27	1.58±0.19	2.86±0.35	0.54	0.98	100

Figure: 2.4 Distance curve of head circumference among the Aka boys of Arunachal Pradesh

Table 2.5: Descriptive statistics of the girth of the bicep among the Aka boys of Arunachal Pradesh

Age in years	*n*	*Mean ±SE*	*SD ±SE*	*CV±SE*	*Growth*		*Growth gradient*
					Absolute	*% per annum*	
9+	32	18.70±0.26	1.47±0.18	7.91±0.98			67.33
10+	33	19.35±0.26	1.51±0.18	7.83±.96	0.65	3.47	69.67
11+	31	20.11±0.27	1.50±0.19	7.49±0.95	0.76	3.92	72.41
12+	31	20.88±0.33	1.85±0.23	8.89±1.12	0.77	3.82	75.18
13+	27	22.86±0.56	2.61±0.35	11.43±1.55	1.98	9.48	82.31
14+	28	24.35±0.36	1.92±0.25	7.92±1.05	1.49	6.51	87.68
15+	26	26.11±0.45	2.34±0.32	8.97±1.24	1.76	7.22	94.02
16+	29	26.72±0.45	2.47±0.32	9.26±1.21	0.61	2.33	96.21
17+	30	27.15±0.26	1.46±0.18	5.39±0.69	0.43	1.60	97.76
18+	33	27.77±0.36	2.12±0.26	7.64±0.94	0.62	2.28	100

Figure 2.5: Distance curve of the girth of bicep of the Aka boys of Arunachal Pradesh

Table 2.6: Descriptive statistics of the girth of the chest of the Aka boys of Arunachal Pradesh

Age in years	*n*	*Mean ±SE*	*SD±SE*	*CV±SE*	*Growth*		*Growth gradient*
					Absolute	*% per annum*	
9+	32	61.97±0.52	2.97±0.37	4.79±0.59			73.64
10+	33	64.35±0.60	3.45±0.42	5.37±0.66	2.38	3.84	76.47
11+	31	65.97±0.61	3.42±0.43	5.19±0.65	1.62	2.51	78.39
12+	31	68.04±0.64	3.61±0.45	5.30±0.67	2.07	3.13	80.85

(Cond...)

13+	27	71.12±1.10	5.73±0.77	8.05±1.09	3.08	4.52	84.51
14+	28	75.38±0.81	4.28±0.57	7.67±1.02	4.26	5.98	89.57
15+	26	80.00±1.21	6.17±0.85	7.72±1.07	4.62	6.12	95.06
16+	29	80.39±0.87	4.72±0.62	5.87±0.77	0.39	0.48	95.53
17+	30	82.92±0.59	3.23±0.47	3.90±0.50	2.53	3.14	98.53
18+	33	84.15±0.62	4.24±0.54	4.24±0.54	1.23	1.48	100

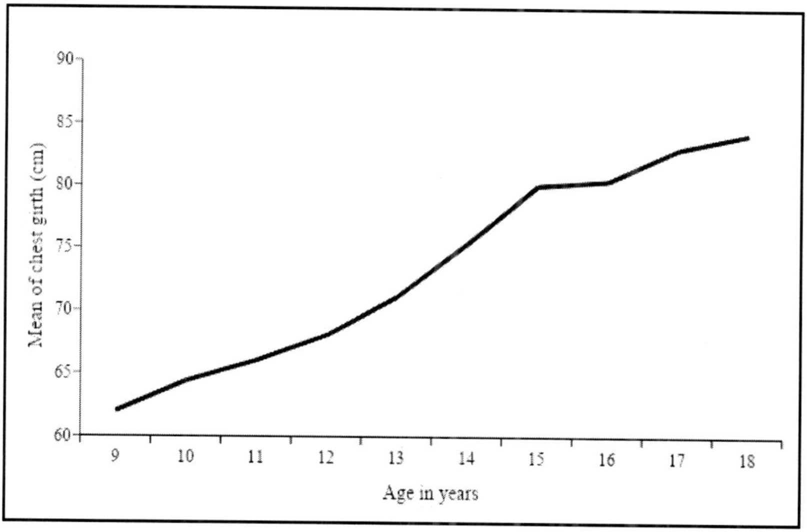

Figure 2.6: Distance curve of the girth of chest among the Aka boys of Arunachal Pradesh

Table 2.7: Descriptive statistics of the girth of the calf of the Aka boys of Arunachal Pradesh

Age in years	*n*	*Mean± SE*	*SD±SE*	*CV±SE*	*Growth*		*Growth gradient*
					Absolute	*% per annum*	
9+	32	24.32±0.24	1.35±0.16	4.79±0.59			72.74
10+	33	25.57±0.35	2.04±0.25	8.00±0.98	1.25	5.13	76.48
11+	31	26.07±0.32	1.82±0.23	6.99±1.78	0.50	1.95	77.98
12+	31	27.28±0.34	1.89±0.24	6.94±0.88	1.21	4.64	81.60
13+	27	28.90±0.48	2.51±0.34	8.68±1.18	1.62	5.93	86.44
14+	28	30.92±0.44	2.37±0.31	7.67±1.02	2.02	6.98	92.49
15+	26	32.18±0.44	2.24±0.31	6.98±0.96	1.26	4.07	96.26
16+	29	32.22±0.37	2.04±0.26	6.34±0.83	0.04	0.12	96.38
17+	30	33.34±0.35	1.93±0.24	5.79±0.74	1.12	3.47	99.73
18+	33	33.43±0.24	1.40±0.17	4.21±0.51	0.09	0.26	100

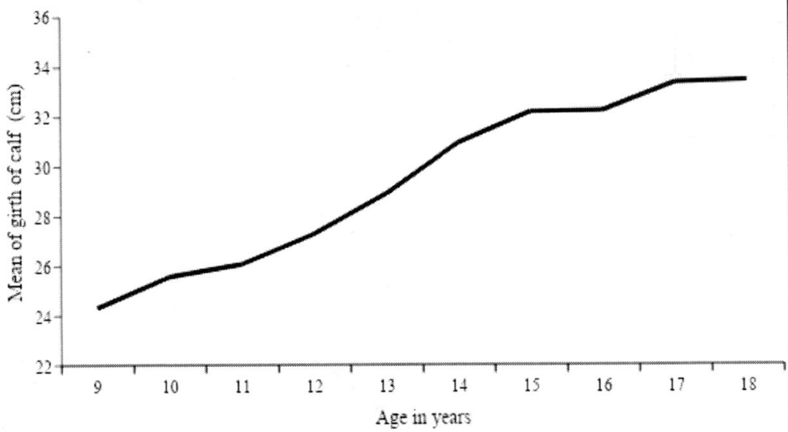

Figure 2.7: Distance curve of the girth of a calf among the Aka boys of Arunachal Pradesh

Table 2.8: Descriptive statistics of body weight of the Aka boys of Arunachal Pradesh

Age in years	*n*	*Mean ±SE*	*SD±SE*	*CV±SE*	*Growth*		*Growth gradient*
					Absolute	*% per annum*	
9+	32	22.81±0.38	2.19±0.27	9.60±1.20	1.39	6.09	43.53
10+	33	24.20±0.52	3.01±0.37	12.46±1.53			46.18
11+	31	27.54±0.69	3.86±0.49	14.03±1.78	3.34	13.8	52.55
12+	31	30.34±1.39	4.68±0.59	15.42±1.95	2.80	10.16	57.9
13+	27	34.82±1.39	7.23±0.94	20.78±2.82	4.48	14.76	66.45
14+	28	41.51±1.07	5.68±0.75	13.68±1.82	6.69	19.21	79.21
15+	26	48.38±1.27	6.50±0.90	13.44±1.86	6.87	16.55	92.32
16+	29	49.23±1.14	6.18±0.81	12.56±1.64	0.85	1.75	93.95
17+	30	52.21±0.80	4.39±0.56	8.41±01.08	2.98	6.05	99.63
18+	33	52.40±0.72	4.14±0.51	7.90±00.97	0.19	0.36	100

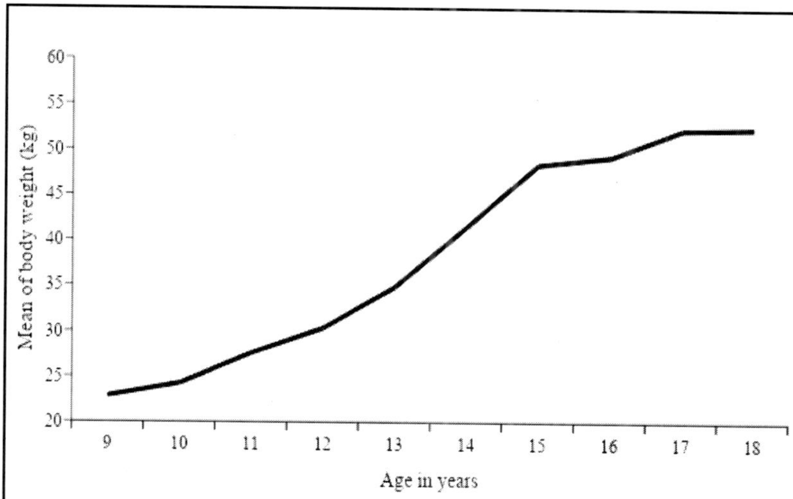

Figure 2.8: Distance curve of body weight of the Aka boys of Arunachal Pradesh

Discussion

WHO defines adolescence as the segment of life between the ages of 10-19 years. They are the in-between group and growth during this stage is faster than at any other time except the first year (Brasel, 1982). Over 80.0 percent of adolescent growth (attained weight and height) is completed in early adolescence (10-15 years), with a marked deceleration in weight and height velocity in the post-pubertal phase (Srikantia, 1989). Among the Aka boys of the present study, more than 90.0 percent of growth is almost complete in all the anthropometric dimensions under study by 15 years of age (when 19 years is taken as the maturity level). In the linear measurements, more than 97.0 percent of maturity status is already achieved in the present study by 15 years of age. This could be because the Aka boys attain puberty in all the linear, circumferential, and ponderal measurements (as evident by the age of highest rate of increment) by 15 years of age. So, maximum growth is already complete. Paula and Dimeglio (2008), calculated the percentage of length reached by the lower limb relative to its final length at different ages from birth upto maturity, at birth it constitutes 20.0 percent of its final length, at puberty, it is 90.0 percent. Among the Aka boys, in circumferential measurements, more growth remains to be completed to achieve the maturity gradient in comparison to the linear and ponderal measurements.

The different parts of the body grow at different rates but the sequence of reaching the adult size differs from one segment to another. In the present study also among the Aka boys, the different body dimensions grow at different rates. The increments between the ages are different for all the measurements under study.

At 9 years the sitting height of the Aka boys has grown only 64.73 percent of the matured size and the lower extremity has already grown 85.94 percent of the matured size. So, the lower extremities have grown more than the sitting height until 9 years of age. The contribution of the lower extremity is, therefore, more to the increase of stature until 9 years. In sitting height, the highest gain is between 13 and 14 years whereas it is between 14 and 15 years in lower extremity length. From the perspective of developmental plasticity, leg length in terms of both absolute size and size relative to height is a sensitive indicator of the quality of life, nutritional status, and socioeconomic environment of children during physical growth.

Among the Aka boys, till the Peak Growth Velocity in sitting height (between 13-14 yrs), 93.2 percent of adult size has been achieved whereas in the lower extremity length till 14-15 years (when it gets the Peak Growth Velocity) the growth is 98.27 percent. Therefore, more growth has taken place in the lower extremity until adolescence and more growth remains to be completed in the trunk length after adolescence. This suggests that the lower extremity grows more whereas lesser growth takes place in the upper extremity contributing respectively to the growth of total body height. This result is being similar to what was analyzed by Tanner (1962), Choudhury (1979), Begum (1994). Tanner has shown that 'in the immediately pre-adolescent years it is the legs which are growing relatively fastest of all skeletal dimensions and so if the body is allowed to grow for an extra two years before the spurt, the legs become relatively longer'. Marshall (1978) even says that the longer period of pre-adolescent growth in boys is also largely responsible for the fact that men's legs are longer than women's concerning the length of the trunk because the legs grow relatively faster than the trunk immediately before adolescence. As per Eveleth and Tanner (1990), Attallah (2019), there is a great variation in body shape, size, and trunk- limb proportion between different populations of mankind, as between Africans, Europeans and Japanese, even individuals of the same height vary widely in their genetically determined trunk – limb proportions. Africans have longer arms and legs relative to the trunk, narrow hips, and more slender calves. The accurate assessment of the physical growth of children has found increasing interests in many fields, such as pediatrics, public health, sports, physical education, ergonomics, and garment industry.

Head circumference measure is at all ages growing in advance than the other body dimensions. Post-natal growth of the human head progresses most intensely in two periods: birth to 7 years and from pubescence to the middle teens. The acceleration in the growth of the human head during adolescence is most conspicuous in the circumference and length and lasts until the 15th or 16th year (Merkel, 1882; Dokladal, 1959). Among the Aka boys at 9 years, the head circumference has attained 92.75 percent of the matured value. The maximum increase takes place between 13 and 14 years (0.78cms) which could be the Peak Growth Velocity in head circumference occurring at adolescence. The existence of an adolescent growth spurt in head circumference is also strongly supported by Eichorn and Baylay (1962) in a longitudinal study of growth in head circumference from birth through young adulthood. The overall gain in head circumference from 9 years to 18 years in the present study is only 4 cms. Head circumference is such a measurement where the maximum amount of growth is already complete at birth.

The increase in growth of skeletal dimensions and body weight during adolescence is the result of hormonal changes that occur during sexual maturation. The timing, duration, and magnitude of the pubertal growth spurt may vary considerably between populations and between individuals within a population (Hauspie and Roelant, 2012). This stage of growth witnesses an intense increase in the rate of growth in height and weight. Man is the only species that are known to have such an increase in skeletal growth velocity. This intense increase in the body dimensions usually begins on average at 10.0 years for girls and 12.0 years for boys. However, there is a considerable variation between individuals and populations. This intensity and duration of the spurt are on average greater for boys than for girls and this accounts for the average sexual dimorphism. The Aka boys show this high intensity of growth between 13 and 14 years in stature when the increase is 8.41 cms and 6.87 kg for body weight. Growth is by no means completed with the end of adolescence, but, for practical purpose, however, it is useful to have an age at which we may say that growth in stature usually ceases, i.e., after which only some 2.0 percent is added (Tanner, 1978). In the present study, among the Aka boys growth could be said to have virtually stopped in stature at 17 years when the increase to the next year is only 0.66cms. Growth in sitting height and the lower extremity length after 17 years is only 0.41cms and 0.25 cms respectively. The changes in circumferential and weight measures continue after adulthood. The overall gain in body weight over ten years is 29.59 kg. Bodyweight is a measurement that reflects any change in any part of the body either increases or decreases.

References

Adolescence- World Health Organization https://www. who.int> documents> pdfs>9241591269_op_handout (Accessed on 21/12/2019)

Attallah, NL 2019, Growth of the limbs and their segments during childhood and adolescence: a photogrammetric study. *Advanced Research Journal of Public Health and Epidemiology,* vol.4, no.2, pp.38-64.

Begum, G 1994, *Age changes in some somatometric characters with reference to nutritional status of Assamese Muslims of Kamrup district, Assam.* Ph.D. Thesis, Guwahati University.

Brasel, J 1982, Changes in body composition during adolescence. In: *Adolescent Nutrition,* M.Winick (Ed.). John Wiley and Sons, New York.

Choudhury, B 1979, *A cross-sectional study of growth and some somatometric characters of Rabha boys of Kamrup, Assam.* Ph.D. Thesis, Guwahati University.

Dokladal Milan.1959. Growth of the main head dimensions from birth up to twenty years of age in Czechs, *Human Biology*, vol.31, no.1, pp. 90-109.

Eichorn, H & Baylay, N. 1962 1962, 'Growth in head circumference from birth through young adulthood', *Child Development*, vol.33, no.2, pp.257-271.

Eveleth, PB, Tanner, JM, 1990, *Worldwide Variation in Human Growth*, (2nd edition), Cambridge University Press.

Hauspie R & Roelant M., 2012. Adolescent growth. In *Human Growth and Development*, B. Bogin (Ed), pp. 57 – 79, Academic Press, Boston.

Marshall, WA 1978, Puberty. In: *Human Growth, 2: Postnatal Growth*, edited by F. Falkner and JM Tanner, New York: Plenum Press: 141-181.

Merkel, F 1882, Beiträge Zur Kenntnis post-embryonaler Entwicklung des menschlichen Schädels. *Festschrift für J. Henle.*

Paula, M Kelly, Alain, Dimeglio 2008, Lower limb growth: how predictable are predictions? *Journal of children's Orthopaedics*, vol.2, no.6, pp.407-415.

Rindfuss, R 1991, 'The young adult years: Diversity, structural change, and fertility,' *Demography*, vol.28, pp.493-512.

Srikantia, SG 1989, *Adolescent Nutrition*- World Health Organization (Accessed on 21/12/19)

Tanner, JM 1962, *Growth at Adolescence* (2nd Edn.) Blackwell Scientific Publication, Oxford.

Tanner, JM 1978, *Foetus into man: Physical growth from conception to maturity.* Open Books, London.

Weiner, JS & Lourie, JA 1981, *Practical human biology*. London: Academic Press.

World Health Organization 2006, Adolescent Nutrition: Review of the Situation in Selected South-East Asian Countries, *World Health Organization, Regional* Office for South-east Asia, New Delhi.

Chapter - 3

Operation of the Determinants of Healthy Ageing in a Tribal Context: A Study among the Karbis of Morigaon District, Assam

CHANDANA SARMAH

Introduction

Demographic ageing is taking place in the Indian population as is evident from the census records over time. India being a vast and regionally diverse country, there is variation in the demographic ageing process in different regions. There are some places like the southern states which have a relatively higher proportion of the elderly and there are states like Assam and Delhi which have a lower proportion of elderly. The states of the country are also culturally heterogeneous and as such the phenomenon, problems, and experiences of the ageing population also differ from region to region.

Assam is a state of the North-eastern part of India and is a multicultural state. There are several scheduled tribes, scheduled castes, and backward castes in addition to other general caste populations inhabiting the state. When we look at ageing from a cultural point of view, we can say that individuals do not just grow old, but they grow old in a social and cultural context. The socio-cultural factors or forces provide the overall context in which an individual develops (Cavanaugh and Fields, 2006). Culture shapes the core beliefs in people and also provides status, social setting, and living conditions for people of all ages. Social institutions organize social relationships and determine roles and functions which in turn formulates the social life of its members. The culture and traditions of society likewise, determine how the given society views older people and the ageing process.

The social institutions determine the status and role of individuals in the community and ensure social integration or isolation. Among a rural tribal community following the traditional mode of occupation, the status of the elderly is high and also ensures a process of healthy ageing.

Healthy ageing is a concept that has been developed by the World Health Organisation in 2015 and it defines it as a process of developing and maintaining the functional ability that enables well-being in older ages. The World Health Organisation has undertaken efforts to operationalize the definition of healthy ageing to construct well-defined public health actions. Healthy ageing comprises functional abilities, intrinsic capacity being the outcome of physical and mental capacities of older people, the impact of the social and physical environment, and well-being. The definition is given by WHO is synonymous with an individual's perception of his/her life in the context of the culture and value system in which he lives. The formulation of this concept over the year has emerged from concepts like 'active ageing', 'successful ageing' and 'productive ageing', etc., which more or less defines similar ideas. In all these different ideas there are a few aspects that are central to the definition of ageing in good health (Cosco *et al.,* 2014). The first being the physiological construct which is identified by physical functioning, the second being engagement in voluntary work, and the third being well-being which is measured from life satisfaction. There are two other aspects which are personal resources and extrinsic factors like finance, which have been used to some lesser extent. Based on the various definitions given by different scholars, overtime two theoretical perspectives have emerged. They are the bio-medical model and the psycho-social dimension. The former stresses physical health, functional and cognitive capacity, eventually supported by psychological dimension and social activity. The psycho-social dimension takes into account well-being and meaningful social activities performed by older people, social participation in different social networks, seeking new opportunities to enjoy a good quality of life in older age.

For this chapter, the classical definition of ageing in good health, referring to the concept of 'successful ageing', created by Rowe and Kahn (1997) is used to assess healthy ageing in a tribal context. Ageing in good health according to this concept is based on a balance between three components. They are the absence of disease and disease-related disability, high functional capacity, and active engagement with life. The paper tries to assess healthy ageing in a tribal context based on the above-mentioned parameters.

Literature review

Studies on ageing in the Indian context are relatively new but the magnitude of literature on various aspects of ageing has over the years grown exponentially. However, concepts have not been framed to suit the Indian context and as such, there is dependence on concepts emerging from the context of developed countries. An understanding of Indian situations will be beneficial for planning developmental actions and for framing concepts based on local issues and contexts.

In the book '*A Life Course Approach to Healthy Ageing*', Kuh *et al.,* (2014) include the themes of healthy ageing as the epidemiology of ageing, physical capability, cognitive capability, psychological and social well-being in addition to outlining the methodological aspects and types of changes that take place in individuals. Kuh *et al.,* (2014) further emphasize that increasing the proportion of healthy and active elder people who remain independent for longer is a way of relieving the costs of an ageing population. According to the authors healthy biological ageing includes three main hierarchical and dynamic components. First is the survival of old age, second is a delay in the onset of chronic disease or disability and third is optimal functioning for a maximal period of life.

Peel *et al.,* (2004) reviewed existing literature which defined and measured healthy ageing from studies on population-based settings from health and gerontology database. The review shows the need to establish a standard for defining and quantifying the concept of healthy ageing. Despite differences, there was a consensus in the studies that the multi-dimensional, positive health outcome should measure the capacity to function well and adapt to environmental challenges in domains assessing physical, mental, and social well-being. Lara *et al.,* (2013) tried to identify Healthy Ageing Phenotypes and the development tools for measuring these features. Given the biological complexity of the aging process, there is no single, simple, and reliable measure of how healthy someone is ageing. After a comprehensive assessment of literature the domains physiological and metabolic health, physical capability, cognitive function, social well-being, and psychological well-being are found to provide a reasonably holistic characterization of the Healthy Ageing Phenotype.

Sowa *et al.,* (2016) tried to look into the predictors of healthy ageing across European countries from various databases. The dependent variable of healthy ageing is constructed as a binary variable with a reference to three dimensions that have been identified in the overview of healthy ageing definition. They are health status self-assessment, functional capacity, and assessment of the perceived meaning of life. Cosco *et al.,* (2014) in the absence of a consensus definition of successful ageing a systematic review across various databases was made. They found that the predominant variable was physical functioning, followed by engagement in voluntary work, life satisfaction, personal resources, and finance.

Goswami and Bhattacharjee (2017) in their study on India's challenges to active ageing in the context of globalization try to determine the challenges and suggest solutions to them. According to them, the main challenges are the huge size of the elderly population, more female elderly, and most elderly living in rural areas. There is also the shift from communicable to non-communicable diseases among them in addition to having economic and social insecurity. Gender inequality, psycho-social problems, and improper nutrition are other challenges. Nair and Thara (2019) undertook a study on active and healthy ageing among scheduled tribes in India from secondary sources. Employing the McGahan Composite Index normalization the composite index was calculated. The health index indicates that one in ten elderly has good health. Community participation Index depicts a better scenario with more elderly scoring moderate and good participation index. The security index shows good among the medium wealth quintile category of the elderly. Overall the scheduled tribe population shows a moderate level of active ageing.

As understood from the literature on healthy ageing, a few aspects remain central to healthy ageing. They are health status, functional capacity, self-assessment of health status, and social well-being. Not much literature is available on the micro-level empirical study of healthy ageing in a community context in India. The present study aims to study healthy ageing among a marginalized scheduled tribe from the Morigaon district of Assam.

Material and methodology

The sample of the study consists of 508 elderly persons in the age group of 60 years and above. To meet the desired sample size data have been collected from twenty villages from the five revenue circles of Laharighat, Mayong, Marigaon, Mikirbheta, and Bhurgaon of the Morigaon district of Assam. All persons who belong to the inclusive age criteria and consented to be part of the study were included. Initially a few hesitated but eventually, all persons in the villages expressed their desire to be part of the study. The composition of the villages is not the same. Some of the villages are inhabited by a single community and a few are of the mixed population like Tiwa, Bodo, Assamese, Koch Rajbangshi, Ahom, etc. eleven villages namely, Bhokotobori, Bongalbori, Majorbori, Burhagaon, Bongaldhora, Majmikir, Rupaibori, Apakarbi, Saligaon, Mantabori, and Posolaghat are composed of only Karbi people. The other remaining 9 villages namely Borigaon, Khorotibori, Bishnupur, Noagaon, Kothalguri, Bherbheri, Tikhabori, Gormari, Sarubori are domiciled by a mixed population, from which only those belonging to the Karbi community were selected.

The components of healthy ageing that have been taken into consideration in this paper are the prevalence of chronic diseases and disease-related disability, high functional capacity, and active engagement with life. To understand the prevalence of chronic diseases, self-reported diseases, as well as those which have been diagnosed by doctors at some point, have been taken into account. Regular consultation with doctors is practically absent and in case of difficulty in managing a health situation by home remedies, they are found to consult ethnomedical practitioners. Therefore, self-reported diseases recognized based on symptoms had to be considered. Disability arising out of diseases has been looked into from bed or work disability arising out of disease conditions whether temporarily or permanently. To understand the health condition more objectively, the nutritional status of the elderly persons was also assessed based on the Body Mass Index (BMI). Body Mass Index is calculated from stature and weight using the standardized tools of anthropometric rod and weighing scale. The elderly persons have been classified according to the WHO classification of nutritional grading for the adult Asian population. Individuals who were unable to stand erect were excluded from the assessment of BMI. The objective was to understand the risk for metabolic complicacy arising out of obesity as well as nutritional complicacies arising out of under-nutrition.

Functional capacity has been examined from the activities of daily living. Both personal care Activities of Daily Living (ADL) and Instrumental Activities of Daily Living (IADL) were taken into consideration. ADL included chewing, bathing, arising, and walking. The elderly persons were asked to assess their ability as facing ‘no difficulty’ or ‘some difficulty’. The IADL includes activities like independently moving out of home, bladder control, preparing meals, and shopping. For the IADL the elderly were asked to assess their ability in three grades. They are ‘no difficulty’, ‘some difficulty’, and ‘need assistance’.

Active engagement with life has been assessed from their current activity status. For this purpose, the activity status of elderly persons has been divided into five categories. They have continued involvement in cultivation or any other economically gainful activity, actively involved in social activity with or without involvement in economic activity, continued active involvement in the household activity, limited involvement in household activity due to physical limitations with age, and those who are only able to take care of their personal care needs or in other words, they are unable to remain engaged in any other familial or social role.

Another dimension integral to healthy ageing is well-being. To have an understanding of the elder person's perception of health, self-assessed health has been taken into account. The elderly persons were asked to assess their health as either 'good', 'fair', or 'poor'. Self-assessed health is taken as an important indicator of well-being as well as mental health.

Findings

A brief description of the basic demographic characteristics is given in table 3.1.

Table 3.1: Distribution of the elderly Karbis by their social variables

Age sex distribution					
Age groups	*60-69 years*	*70-79 years*	*80 years and above*		*Total*
Male	185 (73.12)	51 (20.16)	17 (6.72)		253 (100)
Female	215 (84.31)	29 (11.37)	11 (4.13)		255 (100)
Total	400 (78.74)	80 (15.75)	24 (5.51)		508
Living Arrangement					
	Living alone	*With only spouse*	*Nuclear family*	*Extended family*	*Total*
Male	4 (1.58)	10 (3.95)	35 (13.83)	204 (80.63)	253
Female	2 (0.78)	4 (1.57)	32 (12.55)	217 (85.10)	255
Total	6 (1.18)	14 (2.75)	67 (13.19)	421 (82.87)	508
Educational level					
	Non-literate	*Primary level*	*Middle school*	*High school and above*	*Total*
Male	151 (59.68)	62 (24.51)	11 (4.35)	29 (11.46)	253
Female	223 (87.45)	22 (8.63)	4 (1.57)	6 (2.35)	255
Total	374 (73.62)	84 (16.53)	15 (2.95)	35 (6.89)	508

(Cond...)

Marital status					
	Currently married	***Widow/ widower***		***Never married***	***Total***
Male	199 (78.66)	50 (19.76)		4 (1.58)	253
Female	135 (52.94)	114 (44.71)		6 (2.35)	255
Total	334 (65.75)	164 (32.28)		10 (1.97)	508
Current occupation					
	Agriculture	***Wage earning***	***Petty trade/ service***	***No longer involved in economic activity***	***Total***
Male	178 (70.36)	3 (1.19)	3 (1.19)	69 (27.28)	253
Female	153 (60.0)	11 (4.31)	2 (0.78)	89 (34.90)	255
Total	331 (65.16)	14 (2.76)	5 (0.98)	158 (31.10)	508

(Figures in parentheses indicate percentage) Source: A field study

The sample has been divided into three age groups of 60-69 years, 70-79 years, and 80 years and above age group. The majority of the elderly persons in the district belong to the 60-69 years age group (78.74 %). With an increase in age, their proportions come down reaching 15.75 percent in the 70-79 years age group and 5.51percent in the 80 years and above age group. There is some amount of gender difference in the age distribution. The proportion of female elderly is higher in the 60-69 years age group and is lesser than the male elderly in the higher ages. However, the gender difference in age distribution is statistically significant (9.578 at $P < .05$). Joint or extended family is the norm (82.87 %). At least one adult child continues to live with ageing parents out of filial piety. The proportion of elderly persons living alone (1.18 %) or with only spouses (2.75 %) is low. 13.19 percent live in nuclear families as their children have not yet married or have set up independent homes.

The elderly of the Karbi tribe in the Morigaon district is predominantly non-literate. Illiteracy is prevalent among 73.62 percent of the elderly. Only 6.89 percent have studied at the high school level. Among the elderly male Karbis, 78.66 percent have a surviving spouse and in the case of elderly females, the figure is 52.94 percent. 44.71 percent of female elderly no longer have a surviving spouse and 2.35 percent have remained unmarried. The corresponding figure for the male elderly is 19.76 percent and 1.58 percent. The current occupation of the elderly person's show that 31.10 percent are no longer involved in economically gainful activity. Among them, the proportion of female elderly is higher in comparison to the male elderly. Agriculture is their primary occupation and they continue to work in agricultural fields as long as their physical conditions allow them. 65.16 percent of the elderly Karbis continue to work in agricultural fields and the proportion of males is higher in this category than females. Some even continue to be involved in wage-earning (2.75 %) or petty trade or service (0.98 %).

Prevalence of chronic diseases: 58.50 percent of elderly males and 71.37 percent of elderly females did not reportedly have any chronic health problem. 11.07 percent of males and 4.31 percent of females said they have a problem of high blood pressure, which has been diagnosed by a doctor. But in most cases, doctor's prescriptions are not followed. Ailments relating to the digestive system are found to be prevalent among 6.72 percent males and 7.45 percent females.

Table 3.2: Distribution of the elderly Karbis by age, sex, and self-reported chronic conditions

Age group	*60-69 years*		*70-79 years*		*80+ years*		*Total*	
Disease condition	***Male***	***Female***	***Male***	***Female***	***Male***	***Female***	***Male***	***Female***
No chronic ailment	124 (67.03)	165 (76.74)	19 (37.25)	11 (37.93)	5 (29.41)	6 (54.55)	148 (58.50)	182 (71.37)
High blood pressure	19 (10.27)	10 (4.65)	5 (9.80)	-	4 (23.53)	1 (9.09)	28 (11.07)	11 (4.31)
Diseases of Digestive system	10 (5.41)	13 (6.05)	5 (9.80)	5 (17.24)	2 (11.76)	1 (9.09)	17 (6.72)	19 (7.45)
Body pain	25 (13.51)	17 (7.91)	13 (25.49)	9 (31.03)	3 (17.65)	1 (9.09)	41 (16.21)	27 (10.59)
Diseases relating to respiratory tract	1 (0.54)	5 (2.33)	3 (5.88)	1 (3.45)	2 (11.76)	-	6 (2.37)	6 (2.35)
Diabetes	-	-	2 (3.92)	-	-	-	2 (0.79)	-
Weakness	2 (1.08)	2 (0.93)	1 (1.96)	1 (3.45)	1 (5.88)	-	4 (1.58)	3 (1.18)
Others	4 (2.16)	3 (1.40)	3 (5.88)	2 (6.90)	-	2 (18.18)	7 (2.77)	7 (2.75)
Total	185 (100)	215 (100)	51 (100)	29 (100)	17 (100)	11 (100)	253	255

(Figures in parentheses indicate percentage) Source: A field study

Body pain is a major problem found to be prevalent among the elderly. 16.21 percent of males and 10.59 percent of females reported suffering regularly from body pain. Diseases relating to the respiratory tract are found to be prevalent among 2.37 percent males and 2.35 percent females. Diabetes has been reported by 0.79 percent of males. Since the prevalence of this disease needs a diagnosis of a doctor, the prevalence rate may be found to be low. Some others have reported having problems with body pain along with digestive disorders, high blood pressure, or respiratory diseases. The frequency of these individuals is low and as such has been placed in the category – others. The elderly people generally do not consult doctors and do so only when their traditional management system fails to give them relief. In most cases, therefore, health conditions may remain under-reported. Less reporting of prevalent disease points to the elderly considering themselves to be relatively less ill.

The prevalence of high blood pressure and disease of the respiratory tract show some association with age. Body pain has been reported by the elderly of all three age groups. A higher proportion of female elderly did not reportedly have any chronic ailment in comparison to the male elderly. This could be because the female elderly may be consulting medical practitioners less than the male elderly. High blood pressure and body pain are reported more by the male elderly. Digestive disorder prevalence is almost similar in male and female elderly.

The prevalence of disability arising out of disease conditions is practically absent in the community. No elderly reported having any disease relating to disability.

Nutritional status: The nutritional status has been assessed from their Body Mass Index and classified according to James *et al.,* (1988) classification for adult people.

Table 3.3: Distribution of the elderly Karbis by age, sex, and nutritional status

Nutritional status	*Underweight BMI<18.5*		*Normal BMI 18.5-24.9*		*Overweight BMI >25.0*		*Total*	
Age group	*Male*	*Female*	*Male*	*Female*	*Male*	*Female*	*Male*	*female*
60-69 years	42 (22.70)	48 (22.33)	128 (69.19)	144 (66.98)	15 (8.11)	23 (10.70)	185	215
70-79 years	31 (60.78)	17 (58.62)	19 (37.25)	12 (41.38)	1 (1.96)	-	51	29
80+ years	7 (41.18)	5 (50)	9 (52.94)	5 (50)	1 (5.88)		17	10
Total	80 (31.62)	70 (27.56)	156 (61.66)	161 (63.39)	17 (6.72)	23 (9.06)	253	254*

(Figures in parentheses indicate percentage) * 1 elderly female could not be measured. Source: Field study)

31.62 percent of the elderly males and 27.56 percent of the elderly females are under-weight as per their Body Mass Index. These people may be at risk for nutritional deficiency and immunity-related complicacies in their age. Most of the elderly males (61.66 %) and females (63.39 %) show normal nutritional status. The prevalence of over-weight is low. Only

6.72 percent of males and 9.06 percent of females are over-weight. These people are at risk for metabolic complications arising out of over-weight. Prevalence of under-weight shows variation by age and sex. Under-nutrition is higher among male elderly and shows an increase with an increase in age. The prevalence is relatively lower in the 60-69 years age group.

Functional Ability: Functional ability has been examined from personal care Activities of Daily Living and Instrumental Activities of Daily Living. Personal care ADL includes chewing, bathing, arising, and walking. The decline in these activities shows an association with age. The majority of the elderly of the study can perform these activities without any difficulty.

Table 3.4: Distribution of elderly Karbis by the level of ability in personal care activities of daily living

Age group	*Grade of difficulty*	*60-69 years*		*70-79 years*		*80+ years*		*Total*	
		Male	*Female*	*Male*	*Female*	*Male*	*Female*	*Male*	*female*
Chewing	No difficulty	174 (94.05)	201 (93.49)	29 (56.86)	9 (31.03)	5 (29.41)	2 (18.18)	208 (82.21)	212 (83.14)
	Some difficulty	11 (5.95)	14 (6.51)	22 (43.14)	20 (68.97)	12 (70.59)	9 (81.82)	44 (17.79)	43 (16.86)
Bathing	No difficulty	184 (99.46)	214 (99.53)	51 (100)	27 (93.10)	15 (88.24)	10 (90.91)	250 (98.81)	251 (98.43)
	Some difficulty	1 (0.54)	1 (0.47)	-	2 (6.90)	2 (11.76)	1 (9.09)	3 (1.19)	4 (1.58)
Arising	No difficulty	184 (99.46)	214 (99.53)	49 (96.08)	26 (89.66)	14 (82.35)	8 (72.73)	247 (97.63)	248 (97.25)
	Some difficulty	1 (0.54)	1 (0.47)	2 (3.92)	3 (10.34)	3 (17.65)	3 (27.27)	6 (2.37)	7 (2.75)
Walking	No difficulty	179 (96.76)	203 (94.42)	41 (80.39)	16 (55.17)	5 (29.41)	2 (18.18)	225 (88.93)	221 (86.67)
	Some difficulty	6 (3.24)	12 (5.58)	10 (19.61)	13 (44.83)	12 (70.59)	9 (81.82)	28 (11.07)	34 (13.33)

(Figures in parentheses indicate percentage) Source: Field study)

In the case of chewing, most of the elderly 60-69 years of age group did not reportedly feel any difficulty. However, in the subsequent age group, there is a sharp rise to 43.14 percent in males and 68.97 percent in females. This increases to 70.59 percent in males and 81.82 percent in females in the 80 years and above age group. As for bathing, not much difficulty is found. Only 6.90 percent of females of the 70-79 years age group and 11.76 percent males and 9.09 percent of females of the 80 years and above age group reported facing difficulty in bathing. Arising from a sitting position is another difficulty faced by elderly persons. Elderly from the 60-69 years did not face much difficulty in arising but their proportions increases in the subsequent ages, reaching 17.65 percent in males and 27.27 percent in females of the 80 years and above age group. The elderly of the

80 years and above age group mostly reported facing difficulty in walking. In the case of female elderly, the proportion is low (5.58 %) in the 60-69 years age group, and rises to 44.83 percent in the 70-79 years age group and to 81.82 percent in the highest age group. These elderly reported facing difficulty in performing these activities with an increase in age. However, despite the difficulties, they carry on the activities with reduced ability.

Instrumental Activities of Daily Living are significant for maintaining independent living among elderly persons.

Table 3.5: Distribution of the elderly by the ability to perform instrumental activities of daily living

Age group	*Grade of difficulty*	*60-69 years*		*70-79 years*		*80+ years*		*Total*	
		Male	*Female*	*Male*	*Female*	*Male*	*Female*	*Male*	*female*
Indepen-dently Moving out	No difficulty	175 (94.59)	179 (83.26)	24 (47.06)	4 (13.79)	2 (11.76)	1 (9.09)	201 (79.45)	184 (72.16)
	Some difficulty	10 (5.41)	34 (15.81)	26 (50.98)	21 (72.41)	12 (70.59)	7 (63.64)	48 (18.97)	62 (24.31)
	Need assistance	-	2 (0.93)	1 (1.96)	4 (13.79)	3 (17.65)	3 (27.27)	4 (1.58)	9 (3.53)
Bladder control	No difficulty	183 (98.92)	199 (92.56)	47 (92.16)	3 (10.34)	12 (70.59)	3 (37.50)	242 (95.65)	205 (80.39)
	Some difficulty	2 (1.08)	16 (7.44)	4 (7.84)	26 (89.66)	5 (29.41)	8 (72.73)	11 (4.35)	50 (19.6)
	Need assistance	-	-	-	-	-	-	-	-
Preparing meals	No difficulty	168 (90.81)	205 (95.35)	21 (41.18)	12 (41.38)	4 (23.53)	3 (27.27)	193 (76.28)	220 (86.27)
	Some difficulty	17 (9.19)	5 (2.33)	29 (56.86)	15 (51.72)	11 (64.71)	16 (54.55)	57 (22.53)	26 (10.20)
	Need assistance	-	5 (2.33)	1 (1.96)	2 (6.90)	2 (11.76)	2 (18.18)	3 (1.19)	9 (3.53)
Shopping	No difficulty	180 (97.30)	187 (86.98)	33 (64.71)	4 (13.79)	9 (52.94)	1 (9.09)	222 (87.75)	192 (75.29)
	Some difficulty	5 (2.70)	23 (10.70)	17 (33.33)	22 (75.86)	5 (29.41)	7 (63.64)	27 (10.67)	52 (20.39)
	Need assistance	-	5 (2.33)	1 (1.96)	3 (10.34)	3 (17.65)	3 (27.27)	4 (1.58)	11 (4.31)

(Figures in parentheses indicate percentage) Source: A field study

For the study, those activities which have significance in functioning in the cultural context of the Karbis of the current situation have been taken into account. The elderly persons have been asked to rate their ability in three options. Those who reported needing assistance are considered as not having independence in performing these activities. In the ability to independently move about both within and outside the household, 1.58 percent of the elderly males and 3.53 percent of elderly females reportedly needed assistance. The majority of the elderly are capable of independently moving about. 18.97 percent male and 24.31 percent female elderly faced some difficulty. Facing difficulty and needing assistance shows an association with age and more female elderly reported facing difficulty and needing assistance in independently moving about than elderly males. 0.93 percent of female elderly from 60-69 years reported needing assistance which increased to 13.79 percent among women of 70-79 years age group and reached 27.27 percent in the 80 years and above age group. In the case of bladder control, more female elderly reported facing difficulty in comparison to elderly males. There is also an association with age in this difficulty for both male and female elderly.

Preparing meals is another significant activity indicating an ability for independent living. 1.19 percent of male elderly and 3.53 percent of female elderly reported requiring assistance in preparing meals. This is an indication that the majority of the elderly can continue to prepare meals, among them 22.53 percent male and 10.20 percent female elderly reported facing difficulty in preparing meals. Both facing difficulty and need assistance show an association with age. With an increase in age, the proportion of such people increased.

The female elderly reported to both facing difficulty and needing assistance from the 60-69 years age group but the male elderly reported needing assistance from 70-79 years age group. As for shopping 1.58 percent of male elderly and 4.31 percent of female elderly reported needing assistance. 10.67 percent of male elderly and 20.39 percent of female elderly reported facing difficulty. Both categories show an association with age for both male and female elderly. Female elderly reported to being unable from 60-69 years age group while inability in male elderly is reported from the subsequent age group. From the above discussion, it can be said that the majority of elderly persons can independently carry on their instrumental activities of daily living.

Table 3.6: Distribution of the elderly Karbis by age, sex, and self-rated health status

Self-rated health	*Good*		*Fair*		*Poor*		*Total*	
Age group	*Male*	*Female*	*Male*	*Female*	*Male*	*Female*	*Male*	*Female*
60-69 years	163 (88.11)	183 (85.12)	20 (10.81)	26 (12.09)	2 (1.08)	6 (2.79)	185 (100)	215 (100)
70-79 years	25 (49.02)	10 (34.48)	22 (43.14)	12 (41.38)	4 (7.84)	6 (20.69)	51 (100)	29 (100)
80+ years	4 (23.53)	1 (9.09)	9 (52.94)	7 (63.64)	4 (23.53)	3 (27.27)	17 (100)	11 (100)
Total	192 (75.89)	194 (76.08)	51 (20.16)	45 (17.65)	10 (3.95)	15 (5.88)	253 (100)	255 (100)

(Figures in parentheses indicate percentage) Source: A field study

Activity Status: The daily routine of the elderly persons has been looked into to assess their level of activity within and outside the household. 69.17 percent of male elderly, mainly from the 60-69 and 70-79 years age group are still involved in agriculture for economic subsistence. 5.14 percent of the male is also found to be actively involved in social work in society. After the elderly are not able to be actively involved in agricultural work, they do not stay idle. They continue to be occupied with work that does not require their constant involvement. They take care of the kitchen garden, look after domestic animals, and other matters relating to the running of the household. 13.44 percent are involved in household activities and also look after the kitchen garden and domestic animals. The men who can perform these activities are highest in the 70-79 years age group but decline in the next decade. In the 60-69 years age group, their proportion in this category of work is less as they are still actively involved in agricultural work. 9.88 percent are involved in limited household work indicating a decline in ability. Their proportion shows association with age.

The traditional occupation of agriculture permits an individual to continue to be involved as long as it is physically possible for the person to be working. Being part of the process ensures social contact as well as physical activity for an individual. Among the Karbis both males and females are equally involved within the basic gender division of labour. With age, there is a gradual decline in the amount and level of work in agriculture until physical incapacitation set in.

Table 3.7: Distribution of the elderly Karbis by age, sex, and current activity status

Current activity status	*60-69 years*		*70-79 years*		*80 years and above*		*Total*	
	Male	*Female*	*Male*	*Female*	*Male*	*Female*	*Male*	*Female*
Continuing with Economic activity	159 (85.95)	147 (68.37)	13 (25.49)	1 (3.45)	3 (17.65)	-	175 (69.17)	148 (58.04)
Involved in Social activity	4 (2.16)	1 (0.47)	7 (13.73)	2 (6.90)	2 (11.76)	-	13 (5.14)	3 (1.18)
Actively involved with household activity	16 (8.66)	27 (12.56)	15 (29.41)	5 (17.24)	3 (17.65)	-	34 (13.44)	32 (12.55)
Limited Household activity	5 (2.70)	38 (17.67)	13 (25.49)	19 (65.52)	7 (41.18)	9 (81.82)	25 (9.88)	66 (25.88)
Only personal care activity	1 (0.54)	2 (0.93)	3 (5.88)	2 (6.90)	2 (11.76)	2 (18.18)	6 (2.37)	6 (2.35)
Total	185	215	51	29	17	11	253	255

(Figures in parentheses indicate percentage) Source: A field study.

58.04 percent of the female elderly continue to be involved with agriculture and most of these females belong to the 60-69 years age group and their proportion declined sharply in the subsequent age group. None of the women of the 80 years and above age group are involved in such activity. 1.18 percent of the female elderly are involved in social activity and they belong mostly to the 70-79 years age group. 12.55 percent of the female elderly are actively involved in household activities. Their proportion is lesser in the 60-69 years age group but increased in the subsequent age group. This is because, in the youngest age group, they continue to be economically active. The elderly who are involved in limited household activity indicates a decline in their physical capacity. The proportion of such females shows an association with age. Similar is the case for those who can take care of only their care needs.

Healthy ageing among the Karbis: Healthy ageing in this chapter has been examined as has been already mentioned, from the aspects of the absence of disease and disease-related disability, high functional capacity, and active engagement with life.

The absence of disease can be found among 58.50 percent of males and 71.37 percent of female elderly Karbis. In the absence of regular consultation with doctors, self-reported diseases have been taken into account. Disease prevalence is strongly associated with the psychological connotation of being sick and assuming the sick role as emphasized by medical anthropological literature. From this point of view, the majority of the elderly can be said to be in good health from their contextual connotation of health.

In the objective evaluation of nutritional status, it is found that the majority of the male (61.66 %) and female (63.39 %) elderly show normal nutritional status. The elderly persons having undernutrition may be at greater risk of illness, have poorer recovery from illness, and poorer quality of life. Similarly, overweight enhances the probability of risk from metabolic complicacies. The proportion of such elderly is 6.72 percent male and 9.06 percent female among the elderly Karbis. Overweight is higher among females. Thus, it can be said that the majority of the elderly from nutritional status are also in a good state of health. Disability arising out of disease condition is not found.

Functional ability assessed from the personal care activities of daily living (ADL) and instrumental activities of daily living (IADL) shows that the elderly predominantly continue to be functionally active. None of the elderly reported being unable to perform their care ADL. Facing difficulty in chewing is the highest reported difficulty in ADL and can be said to be the result of a general age-related decline in oral health. The second most reported difficulty is walking, but none reported being unable to perform this activity. Very few elderly reported facing difficulty in personal care ADL of bathing and arising. Therefore, it can be said that in the personal care ADL, elderly Karbis predominantly maintained a high level of functional capacity.

In the instrumental activities of daily living (IADL), a small proportion of the elderly reported needing assistance in performing the activities of independently moving out, preparing meals, and shopping.

These elderly can be said to not have a high level of functional capacity. The proportion of elderly facing some difficulty in performing these IADL is much lower than those who do not face any difficulty. Thus, from the ability to carry out the IADL, it can be said that the majority of the elderly can maintain a high level of functional capacity.

In the assessment of active engagement with life, the current activity status of the elderly Karbis has been considered. Less than 3.0 percent of the elderly are no longer able to contribute to either family or remain actively engaged with society or community life. They can take care of only the care activities of daily living. The elderly who have reported being involved in limited household activity face limitations in performing household chores but still contribute within their capacity. The age-related decline in their ability prevents them from being active contributors. If those continuing to be involved in economic activity, social activity, and actively involved in the household activity are taken together, then it can be said that active ageing is found among the majority of the elderly Karbi males and females of Morigaon district.

Conclusion

The social variables of marital status, living arrangement, and current occupation of the elderly Karbis indicate that they are having an active and healthy ageing process through their remaining socially integrated and having a support system. Proportions of widows are high among female elderly, but none of them are found to be living alone. The dominant joint/ extended family structure indicates the elderly continuing to remain a part of the family support structure. Subsequently, the examination of their activity status points to their continued contribution to the family division of labour. Their involvement in the agricultural processor in the economy allows them to scope for social interaction.

Thus, from the evaluation of all the three aspects of healthy ageing, it can be concluded that the elderly Karbis are having a healthy ageing process. To substantiate their quality of life and well-being, the self-assessment of health status also presents a supportive picture.

Acknowledgement

The study is a part of an ICSSR sponsored research project on bio-social aspects of ageing among the Tiwa and Karbi communities of Morigaon District, Assam. The author expresses her gratitude to the organization for financial support and to research assistant Ms. Priyanka Morang and Ms. Franky Boro for collecting the data.

References

Cavanaugh, JC & Fields, FB 2006, *Adult Development, and Ageing* (5th ed.) USA Wadsworth Thompson Learning Inc.

Cosco, TD, Prina, AM. Perales, J & Stephan, BCM 2014, 'Operational definitions of successful ageing: A systematic review'. *International Psychogeriatrics*, Vol. 26, no.3, pp.373-381.

Goswami, S & Bhattacharjee, J 2017, 'India's challenges to active ageing in the context of globalisation and their solutions: A critical analysis', *International Journal of Current Research,* Vol. 9, no.11, pp. 60626-60631.

James, WPT.; Ferro-Luzzi A & Waterlow, JC., 1988. Definition of chronic energy deficiency in Third World adults. *European Journal of Clinical Nutrition*, vol. 42, pp 969-981.

Kuh, D; Cooper, R; Hardy, R; Richards, M & Ben-Shlomo, Y 2014, *A Life Course Approach to Healthy Ageing,* Oxford University Press, Oxford, OX26DP, UK.

Kuh, D, Richards, M, Cooper, R, Hardy, R & Ben-Shlomo, Y 2014, Life course epidemiology, ageing research, and maturing cohort studies: A dynamic combination for understanding healthy ageing in *A Life Course Approach to Healthy Ageing* (Eds) Kuh, D., Cooper, R., Hardy, R., Richards, M., Ben-Shlomo, Y., Oxford University Press, Oxford, OX26DP, UK.

Lara, J, Godfrey, A, Evans, E, Heaven, B, Brown, IJE, Baron, E, Rochester, L, Meyer, TD & Mathers, JC 2013, Towards measurement of the healthy ageing phenotype in lifestyle-based intervention studies, *Maturitas*, vol.76, pp.189-199.

Nair, SB & Thara, ET 2019, 'Active and healthy ageing among scheduled tribes in India'. *Journal of Gerontology and Geriatrics Research*, Vol. 8, no.1, 495. DOI: 10.4172/2167-7182.1000495.

Peel, N, Bartlett, HM & Clure, R 2004, 'Healthy ageing: How is it defined and measured'. *Australasian Journal on Ageing*, Vol. 23, no.3, pp. 115-119.

Rowe, J W & Kahn, RL 1997, *Successful Ageing*, Gerontologist, 37, 433-440.

Sowa, A, Adamczyk, BT, Topor-Madry, R, Poscia, A & Igrazio la Milia, D 2016, 'Predictors of healthy ageing: Public health policy targets'. *BMC Health Service Research* 16 (Suppl 5): 289 DOI10.1186/s12913-016-1520-5.

World Health Organisation 2015, *World Report on Ageing and Health*, Luxembourg.

Chapter - 4

Factors Determining the Choice of Place of Delivery among the Mishing Mothers of Dhemaji, Assam: A Cross-Sectional Analysis

Maitreyee Sharma

Introduction

According to the latest Sample Registration System (SRS-2015-2017) of India, the nation's maternal mortality ratio (MMR) has been declined from 130 per one lakh live birth in 2014-2016 to 122 per one lakh live birth in 2015-2017. But these figures are still high when compared with the developed countries worldwide and India's own neighbouring countries like China, Bhutan, and Sri Lanka. Though the government of India data shows the effect of various schemes such as Janani Suraksha Yojana (JSY), Janani Shishu Suraksha Karyakaram (JSSK) etcetera in increasing the number of institutional delivery in the rural areas but various researches consistently reaffirm the fact that home deliveries persist despite various initiatives in many tribal and rural belts of India in general and North-east India in particular.

Most maternal and neonatal complications arise soon after the delivery (Darmstadt *et al.,* 2009) and post-partum attention is extremely necessary to ensure both maternal and neonatal survival. In Meghalaya, only 51.4 percent of childbearing women approach a health facility for institutional delivery with a huge rural-urban gap (NFHS-4). Whereas in Assam, the NFHS-4 report shows an increase in institutional deliveries and a decline in MMR but a significant proportion of women still prefer to opt for home delivery than that of institutional delivery. The quality of child delivery can be enhanced by implementing and providing antenatal care (ANC) when it is the best opportunity for the health workers to convince the expecting mothers for safe delivery at the health facility. But in India as a whole, it is observed that the mothers, as well as their family members rarely, provide much attention to antenatal care utilization.

The process of childbirth and obstetric issues become more and more complex when these issues lie deep under the cover of the socio-cultural matrix of society. India, being a nation with numerous populations with varied ethnic backgrounds and beliefs, these issues cannot be solved without understanding the cultural values attached to these life processes and functions. Here, child delivery is believed to be a natural process where medical intervention is not deemed necessary (Kesterton *et al.,* 2010) and an unassisted delivery is considered a sign of courage (Montagu *et al.,* 2011). The choice of the delivery place has been consistently been associated with maternal and neonatal outcomes (Kifle *et al.,* 2018) and in India, there are numerous factors unique to other countries and populations which limit and restrict the expecting mothers to approach the health facilities for ANC utilization as well as for institutional delivery. The government of India's initiatives is being implemented by the National Rural Health Mission (NRHM) since 2005 but the objectives and benefits of such schemes are rarely been understood and realized at the grass-root level.

The Study: Objective, respondents, and methodology

This is a cross-sectional study conducted among the Mishing community inhabiting Batuwa village of Dhemaji district, Assam. Batuwa village is situated at the south bank of the river Moridhal, about 18 kilometers away from the Dhemaji sub-division in the Dhemaji district. The present study mainly aims at understanding the factors that determine the choice of delivery place and identifying the socio-demographic and obstetric parameters that are associated with the preference of place of delivery among the Mishing women of childbearing age. The multistage sampling method was used to select 103 women of childbearing age (20-44 years), who delivered a child in the last five (05) years. A semi-structured interview schedule was designed to harness and record the necessary information on the aim of the research (socio-demographic and obstetric parameters). The data are collected mainly through open-ended interviews in the vernacular dialect. Verbal consent was taken from the respondents before collecting information on the objective.

For the Mishings of Batuwa village of Dhemaji, the nearby health center/facility is 12 km away from the village, and in absence of any other communicating facilities; they have to walk to the PHC. The village so selected is homogeneous with the Mishing population. Ethnically, the Mishings are Mongoloid and belong to the Indo-Tibetan linguistic group. Though the Mishings traditionally used to prefer joint families, at present who can afford to build a new house, they prefer to live in nuclear families. This is mainly due to the change in occupational pursuits due to recurrent floods and population expansion leading to fragmentation of landholdings. For household per capita monthly income, an arbitrary classification is generated according to the actual income of the households in the community as follows: above 75th percentile (Rs 4,500) as a high-income group (HIG), 50th percentile (Rs 3,000 - Rs 4,500) as middle-income group (MIG) and below 50th percentile (less than Rs 3,000) as a low-income group (LIG). Data were further assessed by using the bivariate analysis to determine the magnitude of the relationship between the outcome variable and the independent variables if there was any. For this study, a p-value of ≤ 0.05 and a p-value of ≤ 0.01 were considered statistically significant. The data were analyzed with the help of SPSS software (14.0 versions).

Results

Socio-demographic profile of the respondents (the mothers)

A total of 103 eligible women were included who gave all the information as asked. The mean and the median age at marriage were found to be 33.26±4.07 years and 28.0 years respectively. The highest percentage of women fall in the age group of 25-29 years and 79.62 percent of women are found to be young (20-34 years). Regarding the marital status at the time of the study, of 103 women, 93.20 percent were living with their husbands whereas 6.80 percent got separated after the birth of the last child. No instance of a widow is found among the respondents. The separated women fall within the age range of 33-44 years. As regards the educational status of the respondents and their husbands (including the separated husbands), a considerable percent of them are found to be non-literate and the highest percentage of women are found to be literate up to secondary standard followed by primary standard whereas, among the husbands, the highest percentage of them studied up to primary standard followed by secondary standard. The majority of the women and their husbands were found to be engaged in cultivation followed by petty business. A few of them are engaged in skilled work and the case of husbands in other professional jobs. The majority of the respondents fall in the category of the low-income group with household per capita monthly income less than Rs. 3000 followed by the middle-income group and a few households fall in the high-income group. The majority of the households were found to be joint families and the separated respondents were found to be living with maternal relatives.

Table 4.1: Socio-demographic parameters of the respondents in Batuwa village, Dhemaji district

Parameters	*Frequency*	*Percent (%)*
The current age of the mothers 20-24	11	10.68
25-29	39	37.86
30-34	25	24.27
35-39	16	15.53
40+	12	11.65
Mean±SD	33.26±4.07	
Marital status Ever married	92	89.32
Separated	11	10.68
Educational status of the mothers Non-literate	29	28.16
Primary standard	27	26.21
Secondary standard	47	45.63
Educational status of the husbands (n=103) Non-literate	27	26.21
Primary standard	39	37.86
Secondary standard	29	28.16
Matriculate and above	8	7.77
Occupation of the mothers Cultivation	65	63.11
Skilled work	11	10.68
Petty business	27	26.21
Occupation of the husbands (n=103) Cultivation	69	67.0
Skilled work	8	7.77
Petty business	16	15.53
Lower Profession	6	5.83
Middle Profession	4	3.88
Type of family Nuclear	36	34.95
Joint	67	65.05
Household per capita income LIG	61	59.22
MIG	33	32.04
HIG	9	8.74

Place of delivery and attendants present during delivery

In the present study, the majority of the deliveries occurred at home. Here, home delivery is defined as the delivery at home which was not assisted by a trained health worker using a safe delivery kit. The majority of women (79.61%) delivered their last child at home. In the present study, the highest percentages of home deliveries were conducted by traditional birth attendants (mid-wife) followed by the experienced elder female of the household or the neighbour. Here, it has been found that generally, the elder female belonged to the maternal family of the mother if the respondent is separated or she is a senior female in the case of the joint family. A few numbers of deliveries were also conducted by the Accredited Social Health Activist (ASHA) workers as they were suddenly being requested by the husband at the time of delivery. Only 20.39 percent of deliveries occurred at the public health facility and interestingly, the majority of these deliveries were assisted by ASHA workers. Only 28.57 percent of deliveries were assisted by health workers of the public health facility.

Table 4.2: Percentage distribution of mothers who delivered in the last 5 years according to the place of delivery and attendants present during delivery

Delivery characteristics	*Frequency (n=103)*	*Percent (%)*
Place of delivery Home	82	79.61
Public health facility	21	20.39
Person assisted delivery in Public health facility Health workers	6	28.57
ASHA workers	15	71.43
Attendants during home delivery Traditional birth attendant (midwife)	58	70.73
The elder female of the household/neighbour	18	21.95
ASHA workers	6	7.32

Reasons for home delivery and institutional delivery

The respondents stated seven reasons for preferring home delivery instead of delivery at a health facility. It is apparent from Table 4.3 that among 82 mothers, 62.20 percent stated that traditional birth attendants were more reliable than the health professionals at PHC, followed by previous deliveries were at home (58.54 %), long-distance to PHC from home (no vehicle to carry, 52.44%) and no need to leave other children (47.56 %). Reasons for institutional delivery include complications during labour at home, fear of complication at home, and the decision of the family member particularly the husband. Most mothers opted for institutional delivery because of fear of complications (80.95%), followed by complications during labour at home (52.38%) and the decision of the family member (42.86%).

Table 4.3: Reasons for home delivery and institutional delivery

Reasons for home delivery (n=82)	N (%)*
All my previous deliveries were at home	48 (58.54)
Faith on traditional birth attendant	51 (62.20)
Home delivery is easy and convenient	37 (45.12)
Lack of money	21 (25.61)
No need to leave other children	39 (47.56)
PHC is too far (no vehicle to carry)	43 (52.44)
Family members prefer home delivery	31 (37.80)
Reasons for institutional delivery (n=21)	
Complications during labour at home	11 (52.38)
Fear of complication at home	17 (80.95)
The decision of the family member	9 (42.86)

*Multiple responses

Obstetric characteristics of the mothers

7.77 percent of the respondents reported that they were less than 18 years of age at the time of marriage and most of the women got married in the age range of 22-25 years of age. In the present study, the range of age at marriage is found to be 14-32 years. Only a single woman got married at the age of 14 years. 1.94 percent of women conceived at the age of below 18 years and a considerable percentage of women (28.16%) conceived for the first time at the age of 26 years and above. Of 103 women, 21.36 percent never visited PHC for an antenatal check-up. The reasons for never visiting PHC for ANC utilization were: Transportation problem, refusal from the family members to accompany them to the public health facility, no pregnancy complications felt and some of them felt shy to share issues of pregnancy with the male health workers of the public health facility. A few of the mothers were unaware of the benefits of ANC utilization. However, the majority of them cited the transportation problems to the public health facility as the village is located far interior with pathetic road conditions. Though 78.64 percent of women visited PHC for antenatal care, only 19.76 percent of them continued ANC routine visits for four times or more, and 80.24 percent limited visits to one to three times. And here also, they cited transportation problems to the public health facility and no one of the family members encouraged and accompany them for ANC check-ups.

Table 4.4: Percent distribution of mothers according to obstetric characteristics

Parameters	*Frequency (n=103)*	*Percent (%)*
Age at first marriage <18 years	8	7.77
18-21	32	31.07
22-25	46	44.56
26+	17	16.50
Age at first conception <18 years	2	1.94
18-21	32	31.07
22-25	40	38.83
26+	29	28.16
Gravidity Primigravida	19	18.45
Gravida 2-4	53	51.45
Gravida 5+	31	30.10
Parity 1	49	47.57
2-4	37	35.92
5+	17	16.50
No. of mothers never visited PHC for antenatal care	22	21.36
Reasons for not visiting ANC services (n=22)		
Transportation problem	8	36.36
Refusal from the family	5	22.73
No pregnancy complications	4	18.18
Feel shy	3	13.64
Lack of Knowledge	2	9.09
No. of antenatal care visits (n=81)		
<4	65	80.24
≥4	16	19.76

Table 4.5 shows the results of the bivariate analysis between the decision of choosing a birthplace and independent variables of interest. Six variables were found to be significantly associated with the choice of place of delivery. Education and occupation of the mothers show a significant association with the choice of place of delivery. The literate mothers were more likely to choose a health facility for delivery than non-literate mothers. Similarly, the mothers engaged in petty business and skilled workers were more likely to deliver in the health facility compared to mothers who were engaged in cultivation. Mothers who had received and utilized antenatal care (ANC) (OR=2.58, 95% CI= 1.322-5.905) were two times more likely to deliver in

a health facility than the mothers who did not have received and utilized antenatal care. Means of getting to the health facility (OR=18.55, 95% CI= 5.664-63.007) was also found to have a significant association with the choice of place of delivery. Other factors that show significant association in determining the place of delivery in the analysis include household per capita income and the type of family. It is observed that the elderly/senior women in the joint family with the experience of home delivery encourage the other women in the family to opt for home delivery rather than approaching the health facility. Respondents in the nuclear family show higher odds of using a health facility during delivery. Mothers who were included in the high-income group (HIG) were more likely to choose a health care facility than the middle income (MIG) and low-income group (LIG) mothers.

Table 4.5: Bivariate analysis of the relationship between the decision for choosing the place of delivery and selected independent variables

Variables	*Respondents who delivered at the health facility*	*Respondents who delivered at home*	*OR*	*95 % CI*	
				Lower	*Upper*
Age (years)					
20-29[ref]	11	39			
30-39	8	33	1.04	0.550	2.0943
40+	2	10	0.825	0.064	0.992
Marital status					
Ever married [ref]	18	74			
Separated	3	8	1.54	0.685	2.991
Education of the respondents					
Non-literate [ref]	2	27			
Primary standard	4	23	3.25**	1.695	7.015
Secondary and above	15	32	2.67**	1.211	4.204
Occupation of the respondents					
Cultivation	6	62			
Skilled work	3	8	3.88**	1.764	7.895
Petty business	12	15	2.12**	1.106	4.286
Previous childbirth					
No [ref]	9	40			
Yes	13	41	1.42*	0.767	2.876
Antenatal care utilization					
Not received care [ref]	2	20			
received care	17	64	2.58**	1.322	5.905
Means of getting to the health facility					

(Cond...)

Walking [ref]	2	55			
By vehicle	19	27	18.55***	5.564	63.007
Household per capita income					
LIG [ref]	4	57			
MIG	10	23	6.20**	1.762	19.351
HIG	7	2	8.04***	2.533	23.876
Type of family					
Joint [ref]	8	59			
Nuclear	13	23	4.15**	1.063	8.872

*p<0.25;**p<0.05;***p<0.001

Discussion

This study aimed at determining the factors that influence the choice of place of delivery in a rural Mishing community in Assam. It was found that several socio-demographic and obstetric factors have a significant influence on the place of delivery among mothers. The present study reveals that most of the deliveries of the Mishing mothers of the village took place at home with the help of untrained mid-wife followed by the experienced elder female of the household or of the neighbour who used to cut the umbilical cord with the help of an unsterilized knife, blade or a sharpened bamboo strip. Thereafter, either the mother or the midwife uses to tie the navel of the infant with cotton thread without any fear of infection. Such a kind of practice may lead to infection and results in the death of the infant.

For the Mishings of Batuwa village of Dhemaji, the nearby health center is 12 km away from the village, and in absence of any other communicating facilities; they have to walk to the PHC. This is one of the reasons for preferring delivery at home rather than approaching the public health facility. According to the elders of the society, it is their age-old tradition and they lack faith in the health professionals. A few numbers of deliveries were also conducted by ASHA workers as they were suddenly being requested by the husband at the time of delivery and they assisted the deliveries at home due to transportation problems. The prevalence of home delivery was found to be 79.61 percent which is quite higher than the prevalence of home deliveries in rural areas of India (24.9%) according to NFHS-4. However, in Assam, the prevalence of home delivery is found to be 31.8 percent and in Dhemaji district it is found to be 25.4 percent. In the present study, only 20.39 percent of deliveries occurred at the public health facility and interestingly, the majority of these deliveries were assisted by ASHA workers and a few are assisted by the health workers of the public health facility.

All the institutional deliveries in the study are conducted in the public health facility. The respondents stated several reasons for preferring home delivery instead of delivery at a health facility. It is apparent from the statements of the respondent mothers that traditional birth attendants were more reliable to them than the health professionals at the public health facility (Shiferaw *et al.,* 2013). Some of the respondents felt shy to go to the male health workers of the facility. Mothers with previous deliveries at home preferred again to choose the birthplace at home than a public health facility. Other reasons that compelled the mothers to choose home delivery are long-distance to public health facilities from home (no vehicle to carry), lack of money, and no need to leave other children. Similar findings were also found by Balaji *et al.,* (2003), Das *et al.,* (2010), Sudhakaram (2009), and Sarkar *et al.,* (2018). Poor economic condition is one of the important constraints why they do not opt for institutional delivery. Though the majority of the respondents were aware of government schemes such as Janani Suraksha Yojana (JSY) and Janani Shishu Suraksha Karyakaram (JSSK) to lower their cost during delivery at the public health facility as they get the money later and the amount of money received according to them is often less than spent during the travel and lodging expenses of attendants. This finding corroborates with the findings of Sarkar *et al.,* (2018). Reasons for institutional delivery include complications during labour at home, fear of complication at home, and the decision of the family member particularly the husband. A similar finding was observed in other studies also (Sarkar *et al.,* 2018; Kitui *et al.,* 2013) Most mothers opted for institutional delivery because of fear of complications followed by complications during labour at home and the decision of the family member. A considerable percent of respondents never visited PHC for an antenatal check-up. Though 78.64 percent of women visited PHC for antenatal care, only 19.76 percent of them continued ANC routine visits for four times or more, and 80.24 percent limited visits to one to three times. The reasons stated were mainly transportation problems, refusal from the family and no one to accompany, no pregnancy complications, felt shy to communicate with male health workers and a few of the respondents did not understand the need for antenatal care visits.

According to NFHS-4, in India, only 21 percent of pregnant women utilized full ANC. In the Dhemaji district, only 47.3 percent of rural pregnant women opted for at least 4 antenatal care visits and in Assam, the percentage is found to be 44.8 percent. In the Indian context, economic status, no encouragement from family members, and maternal education are highly associated with incomplete ANC utilization (Kumar *et al.,* 2019). The results of the bivariate analysis between the decision of choosing a birthplace and independent variables of interest show that six variables were found to be significantly associated with the choice of place of delivery. The present study reaffirms the findings from different studies that education and occupation of the mothers significantly predict the place of childbirth. Mukhtar *et al.,* (2018) found that women with no educational background are most likely to deliver at home and respondents with a higher level of education are more likely to utilize the healthcare delivery services in Srinagar, India. These findings are found to be consistent with the results of studies conducted by Ravi and Kulasekaran (2014) in Tamil Nadu, India. The studies by Navaneetham and Dharmalingam (2002) and Sahoo *et al.,* (2015) confirm that the education leads to better health awareness, and this may sensitize choices for the quality of healthcare and educated couples may be more open to modern medicine, aware of the benefits of skilled attendants, more able to communicate with health workers and demand appropriate care. Household per capita income and related factors, such as the occupation of the respondents were found to be associated with place of delivery in this study. A similar finding is also apparent in the study made by Sahoo *et al.,* (2015), Bhattacharyya *et al.,* (2016), and Varma *et al.,* (2010). Other researches consistently show that low income and high expenditure in case of institutional delivery are serious obstacles in maternal healthcare utilization (Mohanty and Shrivastava, 2013).

The present study also corroborates with other studies on antenatal care utilization and place of delivery choice. Mothers who had received and utilized antenatal care (ANC) were two times more likely to deliver in a health facility than the mothers who did not have received and utilized antenatal care. Women who had utilized four or above antenatal visits during their last pregnancy chose a health care facility for the childbirth. This may be due to the influence of the advice provided by the health care workers during antenatal care resulting in the increased use of health facility delivery (Mukhtar *et al.,* 2018). This study did not show a significant association between parity and the use of health care facilities for childbirth. However, many studies have found high parity as a predisposing factor for home delivery among the study participants (Mukhtar *et al.,* 2018; Ravi and Kulasekaran, 2014). This study reaffirms that in India, the means of getting to the health facility was also found to have a significant association with the choice of place of delivery. Another contributing factor for home delivery was reported to be a lack of public transport combined with poor roads and the long distance to the nearest health facility performing the delivery. A similar finding is also being reported by Sarkar *et al.,* (2018) in Meghalaya as well as by Sarker *et al.,* (2016) in rural Bangladesh. Gabrysch *et al.,* (2011) showed that the distance to the closest health facility doubled the odds of facility delivery decreased by 29.0 percent. It is observed that the elderly/senior women in the joint family with the experience of home delivery encourage the other women in the family to opt for home delivery rather than approaching the health facility. Respondents in the nuclear family show higher odds of using a health facility during delivery. Other studies also suggest that maternal healthcare behaviour is influenced by the previous reproductive healthcare experience of senior wives in a joint family situation (Jejebhoy and Sathar, 2011; Gorain *et al.,* 2017). Therefore, it is very much important to look into the grass-root level to know the hindrances that do restrict the approach of the rural tribal women to health care facilities and the willingness of mothers and their personnal experiences at health care facilities so that quality improvement programs at the facility can result in increased maternal satisfaction improving the demand for facility births.

References

Balaji, R, Dilip, TR & Duggal, R 2003, 'Utilization and expenditure on delivery care services: some observations from Nashik district, Maharashtra'. *Regional Health Forum*, vol.7, pp.34–41.

Bhattacharyya, S, Srivastava, A, Roy, R & Avan, BI 2016, 'Factors influencing women's preference for health facility deliveries in Jharkhand state, India: A cross-sectional analysis'. *BMC Pregnancy Childbirth*, vol.16, pp.50.

Darmstadt, GL, Lee, AC, Cousens, S, Sibley, L, Bhutta, ZA & Donnay, F 2009, '60 million non-facility births: Who can deliver in community settings to reduce intrapartum-related deaths'? *International Journal of Gynecology & Obstetrics,* vol.107(Suppl 1), pp.89-112.

Das, P, Mausumi, B & Tikadar, T 2010, 'Client satisfaction on maternal and child health services in rural Bengal'. *Indian Journal of Community Medicine*, vol.35, no.4, pp.478-481.

Gabrysch, S, Cousens, S, Cox, J & Campbell, OM 2011, 'The influence of distance and level of care on delivery place in rural Zambia: A study of linked national data in a geographic information system'. *PLoS Medicine*; 8:e1000394.

Gorain, A, Barik, A, Chowdhury, A & Rai, RK 2017, 'Preference in place of delivery among rural Indian women'. *PLoS One*, 12(12): e0190117.

Jejebhoy, SJ & Sathar, ZA 2011, 'Women's autonomy in India and Pakistan: The influence of religion and region'. *Population Development Review*, vol.27, no.4, pp.687-712.

Kesterton, AJ, Cleland, J & Sloggett, A. 2010, 'Institutional delivery in rural India: the relative importance of accessibility and economic status'. *BMC Pregnancy Childbirth,* vol.10, p.30.

Kifle Meron Mehari, HF, Kesete, HT, Gaim Angosom, GS & Araya, MB 2018, 'Health facility or home delivery? Factors influencing the choice of delivery place among mothers living in rural communities of Eritrea'. *Journal of Health, Population, and Nutrition*, vol.37, pp.22.

Kitui, J, Lewis, S & Davey, G 2013, 'Factors influencing place of delivery for women in Kenya: An analysis of the Kenya demographic and health survey, 2008/2009'. *BMC Pregnancy Childbirth*, vol.13, pp.40.

Kumar, G, Choudhary, TS, Srivastava, A & Upadhyay, RP 2019, 'Utilization, equity and determinants of full antenatal care in India: analysis from the National Family Health Survey 4'. *BMC Pregnancy Childbirth,* vol.19, p.327.

Mohanty, SK & Srivastava, A 2013, 'Out-of-pocket expenditure on institutional delivery in India'. *Health Policy and Planning*, pp.1-6.

Montagu, D, Yamey, G & A. Visconti, A. 2011, 'Where do poor women I developing countries give birth? A multi-country analysis of demographic and health survey data'. *PLoS ONE*, vol.6, no.2.

Mukhtar, M, Nilofar, M & Quansar, R 2018, 'Factors influencing the choice of place of delivery among recently delivered women in tribal areas of district Srinagar: A cross-sectional study'. *Journal of Medical Science And clinical Research,* vol.6, no.6, pp. 356-361.

National Family Health Survey 4 Report .2015-2016. *Fact sheet India. Ministry of Health & Family Welfare.* Government of India, International Institute for Population Sciences (IIPS), Mumbai.

Navaneetham, K & Dharmalingam, A, 2002, 'Utilization of maternal health care services in Southern India'. *Social Science & Medicine*, vol.55, no.10, pp.1849–1869.

Ravi, RP & Kulasekaran, RA 2014, 'Does socio-demographic factors influence women's choice of place of delivery in rural areas of Tamilnadu State in India'. *American Journal of Public Health Research*, vol.2, no.3, pp.75-80.

Sahoo, J, Singh, SV & Gupta, VK 2015, 'Do socio-demographic factors still predict the choice of place of delivery: A cross-sectional study in rural North India'? *Journal of Epidemiology and Global Health*, vol.5(4): Supplement 1.

Sample Registration System Data Report 2015-2017, *Office of the Registrar General & Census Commissioner, India.* Ministry of Home Affairs, Government of India.

Sarkar, A, Kharmujai, OM, Lynrah, W & Suokhrie, NU 2018, 'Factors influencing the place of delivery in rural Meghalaya, India: A qualitative study'. *Journal of Family Medicine and Primary Care*, vol.7, pp.98-103.

Sarker, BK, Rahman, M, Rahman, T & Hossain, J 2016, 'Reasons for preference of home delivery with traditional birth attendants (TBAs) in rural Bangladesh: A Qualitative Exploration'. *PLoS One*, 11:e0146161.

Shiferaw, S, Spigt, M & Godefrooi, M 2013, 'Why do women prefer home births in Ethiopia'? *BMC Pregnancy Childbirth*, vol.13:5.

Sudhakaram, V 2009, *Institutional deliveries in India: A socio-economic and cultural view*. http://www.articlesbase.com/womens-health-articles.

Varma, D, Khan, ME & Hazra. ME A. 2010, 'Increasing institutional delivery and Access to emergency obstetric care services in rural Uttar Pradesh'. *The Journal of Family Welfare*, vol.56, pp.23-30.

Chapter - 5

Prevalence and Determinants of Leucorrhea among Dibongiya Deori Women of Assam

Juri Borah and Sarthak Sengupta

Introduction

Vaginal discharge is one of the most common symptoms of gynaecological morbidity (Loretta, 1995). Gynaecological morbidity harms female reproductive ability, mental health, and in performing routine physical activity (Kulkarni and Durge, 2005). Abnormal vaginal discharge or leucorrhea is one of the most commonly reported symptoms of women in developing countries. It is an expression of some underlying disorders which may be either functional or organic. The symptom of vaginal discharge is suffered by the female of any age. It may appear at the age of infancy, childhood, childbearing age, menopause and during senescence (Rakoff and Bland, 1940). The World Health Organization has recommended syndromic management, in which the women suffering from discharge are treated for some or all of the five common reproductive tract infections namely, chlamydia trachomatis infection, gonorrhea, trichomoniasis, bacterial vaginosis, and candidiasis (Patel *et al.,* 2005). Bacterial vaginosis is the commonest cause of abnormal vaginal discharge in the women of childbearing age, but may also be encountered in menopausal women, and is rather rare in children (Schneider *et al.,* 1998; Lamont *et al.,* 2000; Dan *et al.,* 2003; Bhalla *et al.,* 2007).

The community-based studies so far conducted have revealed that women suffer from the problem of leucorrhea silently. The reasons are mentioned to be gender inequalities, cultural restrictions, lack of women autonomy, poor awareness, lack of proper infrastructure, and focused counseling services (Gittlesohn *et al.,* 1994; Oomman, 1996; Bhatia *et al.,* 1997; Wasserheit *et al.,* 1998; Joshi *et al.,* 2008).

Women's perception regarding their problem plays an important role. The severity of the problem is understood by the womenfolk differently which makes it difficult for the researcher to ensure the exact level of prevalence (Kaur and Kapoor, 2014). Earlier studies have shown that women perceive white discharge to be part of sexual maturation, hence, do not concern about it. On the contrary, few women consider abnormal vaginal discharge a non-curable disease. Various studies also observed that women commonly have multiple terms to refer to different types of vaginal discharge and they linked it to other illnesses, like weakness and backache, and used these terms as euphemisms to report discharge (Kanani *et al.,* 1994; Patel, 1994).

Objective

In this chapter, an attempt has been made to understand the influence of socio-demographic variables in the prevalence of abnormal vaginal discharge along with its associated complaints among the Dibongiya Deori women of Assam.

Materials and methods

A total of 504 (rural= 405 and urban= 99) Dibongiya Deori mothers were selected for the study. Selection criteria include the presence of husbands as alive and had at least one surviving child. Respondents from the rural area were residing in exclusively Dibongiya Deori dominated villages namely, Borchapori Deori *Gaon*, Kinapather Deori *Gaon*, Majorchapori Deori *Gaon*, Bhogpur Deori *Gaon,* and Pichola Deori *Gaon*. The urban settlements were Bihpuria, Bandardewa, and Narayanpur. All these areas were in the locality of the Lakhimpur district of Assam.

Deoris are the riverine plains scheduled tribe of Assam. The physical appearance shows affinity with the Mongoloid ethnic stock. As per the 2011 census report, the numeric strength of the Deori population in Assam is 43,750 in number. The Dibongiya Deori may be regarded as the major division of the Deoris. The name Dibongiya owes its origin from the river Dibong of their original homeland Sadiya in Arunachal Pradesh. Deori people are concentrated in various districts of upper Assam namely, Lakhimpur, Dhemaji, Majuli, Jorhat, Dibrugarh, Sibsagar, Tinsukia, and Sonitpur. Some villages are also found in the Guwahati areas of Kamrup district. However, they can be traced from the Lohit and Changlang districts of Arunachal Pradesh too. Lakhimpur district has the highest concentration of Deori population (61.0% of total Deori population of the State) as compared to the other districts. However, the district Lakhimpur consists of a majority of the Dibongiya Deori villages (Department of Welfare of Plains Tribe and Backward Classes, Assam, 2007). They are mainly agriculturists and are the followers of priestly activities. They seem to have preserved their traditional socio-rituals organizations.

Data were collected during the years 2014-16. A pre-tested schedule was applied for data collection. Non-participant observation besides individual as well as group interview methods was adopted for this purpose. Data analysis was performed by the computer software namely, Microsoft Office Excel as well as the Chi-Square Test of Independence.

Results

Table 5.1 shows the occurrence of the problem of vaginal discharge according to the socio-demographic characters of the Dibongiya Deori mothers. Among all the respondents, 22.02 percent were suffering from the problems of vaginal discharge. The proportion of the sufferers from the rural area was 23.46 percent, while it was found to be 16.16 percent in the urban setup. The difference between the rural and urban places of residence was found to be statistically insignificant as per the chi-square test of significance. The prevalence of vaginal discharge was lowest among the mothers aged 50 years and above (9.64%). A comparatively higher proportion of it was recorded among the mothers of the age group of 40-49 years (25.56%), which was immediately followed by the age group of 30-39 years (25.34%). The chi-square test of significance showed statistically significant differences in regards to the prevalence of vaginal discharge as per their present age groups. Respondents with 3 numbers of parity were found to experience the highest occurrence of vaginal discharge (25.93%). Interestingly, the proportion of the sufferers was observed as relatively less among the mothers who had 4 numbers of parity or more than that (19.01%). The respondents who were either graduate or had educational attainment beyond that level (14.29%) experienced relatively less problem with vaginal discharge. The prevalence was proportionately higher among the mothers who were educated up to high school (24.53%), followed by beyond high school level (24.28%) and non-literate (20.00%). Homemakers (22.81%) were found to suffer more than working mothers (14.58%). The respondents from the nuclear family (24.51%) had a relatively higher occurrence of vaginal discharge than the mothers from the joint family

Table 5.1: Vaginal discharge according to socio-demographic characters

Socio-demographic characters		*Respondents*	*Vaginal discharge*		χ^2 *(df), P- value*
			Yes (%)	*No (%)*	
Age group (in years)	≤20-29	142	32 (22.54)	110 (77.46)	9.34 (3), P<0.05
	30-39	146	37 (25.34)	109 (74.66)	
	40-49	133	34 (25.56)	99 (74.44)	
	≥50	83	8 (9.64)	75 (90.36)	
Place of residence	Rural	405	95 (23.46)	310 (76.54)	2.47 (1), P>0.05
	Urban	99	16 (16.16)	83 (83.84)	
Parity	1	139	31 (22.30)	108 (77.70)	1.64 (3), P>0.05
	2	136	29 (21.32)	107 (78.68)	
	3	108	28 (25.93)	80 (74.07)	
	≥4	121	23 (19.01)	98 (80.99)	
Maternal education	Illiterate	80	16 (20.00)	64 (80.00)	3.79 (4), P>0.05
	Primary	57	9 (15.79)	48 (84.21)	
	Up to high school	159	39 (24.53)	120 (75.47)	
	Beyond high school	173	42 (24.28)	131 (75.72)	
	Graduate and above	35	5 (14.29)	30 (85.71)	
Maternal occupation	Homemaker	456	104 (22.81)	352 (77.19)	1.71 (1), P>0.05
	Working	48	7 (14.58)	41 (85.42)	
Family type	Joint	247	48 (19.43)	199 (80.57)	1.89 (1), P>0.05
	Nuclear	257	63 (24.51)	194 (75.49)	
Family planning	Hormonal medicine	124	28 (22.58)	96 (77.42)	6.46 (3), P>0.05
	Sterilized	81	26 (32.10)	55 (67.90)	
	No use	286	54 (18.88)	232 (81.12)	
	Other	13	3 (23.08)	10 (76.92)	
Napkin during menstruation	No use	120	24 (20.00)	96 (80.00)	3.64 (3), P>0.05
	Cloth	240	56 (23.33)	184 (76.67)	
	Disposable	85	14 (16.47)	71 (83.53)	
	Both cloth and disposable	59	17 (28.81)	42 (71.19)	

(19.43%). Quite a good number of Dibongiya Deori mothers experienced the problem of vaginal discharge who had adopted the process of sterilization (32.10%). On the other hand, considerably less prevalence of it was found among those (18.88%) who had not used any family planning devices till the time of the investigation. Considerable proportions of the mothers were found to suffer from the problem of vaginal discharge who used both cloth and disposable as per the requirement (28.81%) during the time of menstruation. The respondents who used only cloths during periods were also found to suffer in greater proportions (23.33%). The chi-square test of significance fails to record any significant differences in regards to parity, maternal education and occupation, family type, family planning, and use of napkin during menstruation.

Symptoms associated with vaginal discharge according to socio-demographic characters among the ever-married mothers are shown in table 5.2. Vaginal discharge was accompanied by the problems of itching (2.78%) and burning sensation (4.96%) at the genitalia, feeling weak (14.09%) and discomfort (6.55%), pain in the lower abdomen (13.10%)

Table 5.2: Symptoms associated with vaginal discharge according to socio-demographic characters

Socio-demographic characters	*Vaginal discharge accompanied by*								*Respondents*
	Itching	*Burning*	*Discomfort*	*Weakness*	*Lower abdominal pain*	*Lower back pain*	*Other problem*	*At least any one of these problems*	
Age group (in years)									
≤20-29	1(0.70)	4(2.82)	8(5.63)	19(13.38)	18(12.68)	6(4.23)	3(2.11)	27(19.01)	142
30-39	5(3.42)	10(6.85)	12(8.22)	24(16.44)	21(14.38)	2(1.37)	8(5.48)	27(18.49)	146
40-49	7(5.26)	10(7.52)	11(8.27)	21(15.79)	24(18.05)	7(5.26)	0	30(22.56)	133
≥50	1(1.20)	1(1.20)	2(2.41)	7 (8.43)	3(3.61)	0	0	6(7.23)	83
Place of residence									
Rural	9(2.22)	19(4.69)	25(6.17)	58(14.32)	54(13.33)	14(3.46)	11(2.72)	76(18.77)	405
Urban	5(5.05)	6(6.06)	8(8.08)	13(13.13)	12(12.12)	1(1.01)	0	14(14.14)	99
Parity									
1	5(3.60)	7(5.04)	7(5.04)	15(10.79)	20(14.39)	3(2.16)	0	27(19.42)	139
2	5(3.68)	4(2.94)	11(8.09)	18(13.24)	14(10.29)	5(3.68)	3(2.21)	21(15.44)	136
3	2(1.85)	8(7.41)	8(7.41)	23(21.30)	15(13.89)	4(3.70)	3(2.78)	22(20.37)	108
≥4	2(1.65)	6(4.96)	7(5.79)	15(12.40)	17(14.05)	3(2.48)	5(4.13)	20(16.53)	121
Maternal education									
Non-literate	1(1.25)	3(3.75)	3(3.75)	12(15.00)	11(13.75)	5(6.25)	2(2.50)	12(15.00)	80
Primary	3(5.26)	3(5.26)	3(5.26)	8(14.04)	5(8.77)	3(5.26)	0	8(14.04)	57
Up to high school	4(2.52)	11(6.92)	9(5.66)	27(16.98)	30(18.87)	3(1.89)	9(5.66)	34(21.38)	159
Beyond high school	4(2.31)	6(3.47)	16(9.25)	22(12.72)	17(9.83)	4(2.31)	0	33(19.08)	173
Graduate and above	2(5.71)	2(5.71)	2(5.71)	2(5.71)	3(8.57)	0	0	3(8.57)	35
Maternal occupation									
Homemaker	13(2.85)	24(5.26)	32(7.02)	65(14.25)	62(13.60)	15(3.29)	11(2.41)	84(18.42)	456
Working	1(2.08)	1(2.08)	1(2.08)	6(12.50)	4(8.33)	0	0	6(12.50)	48
Family type									
Joint	5(2.02)	7(2.83)	8(3.24)	27(10.93)	25(10.12)	3(1.21)	0	37(14.98)	247
Nuclear	9(3.50)	18(7.00)	25(9.73)	44(17.12)	41(15.95)	12(4.67)	11(0.04)	53(20.62)	257
Family planning									
Hormonal medicine	5(4.03)	6(4.84)	7(5.65)	20(16.13)	18(14.52)	3(2.42)	2(1.61)	23(18.55)	124
Sterilized	1(1.23)	7(8.64)	12(14.81)	21(25.93)	18(22.22)	8(9.88)	7(8.64)	21(25.93)	81
No use	7(2.45)	12(4.20)	12(4.20)	30(10.49)	28(9.79)	4(1.40)	2(0.70)	43(15.03)	286
Other	1(7.69)	0	2(15.38)	0	2(15.38)	0	0	3(23.08)	13
Napkin during menstruation									
No use	3(2.50)	7(5.83)	7(5.83)	15(12.50)	16(13.33)	2(1.67)	3(2.50)	19(15.83)	120
Cloth	7(2.92)	12(5.00)	16(6.67)	36(15.00)	34(14.17)	12(5.00)	5(2.08)	44(18.33)	240
Disposable	4(4.71)	4(8.24)	7(8.24)	13(15.29)	13(15.29)	0	3(3.53)	17(20.00)	85
Both cloth and disposable	0	2(5.08)	3(5.08)	7(11.86)	3(5.08)	1(1.69)	0()	10(16.95)	59

and lower back region (2.98%) and other problems (2.18%) include headache and pain at feet. However, 17.86 percent of the respondents had experienced at least any one of these problems. The accompanying problems of vaginal discharge were observed to be comparatively higher in the age group of 40-49 years, followed by the age group of 30-39 years. Among the mothers aged 50 years or later the occurrence of the problems were seemed to be proportionately less. The problem of pain and weakness was suffered as comparatively higher by the respondents from the rural area than their urban counterparts. Contrary to it, the feeling of itching and burning sensation at the genital region and discomfort were reported in higher proportions by the mothers of urban setup. The respondents with 3 numbers of parity were more sufferers with at least any variety of the accompanying problems of vaginal discharge. Working Dibongiya Deori mothers as well as who were graduate or had beyond that level of educational attainment experienced relatively fewer problems than their counterparts. The symptoms associated with vaginal discharge were comparatively more among the respondents residing in a nuclear family. Mothers who did sterilization had suffered relatively more with all these problems except itching in genitalia. The problem of itching was recorded as proportionately higher among those who used other family planning devices such as IUD and condoms to avoid further pregnancy.

Discussion

The prevalence of abnormal vaginal discharge among the Dibongiya Deori mother was quite high (22.02%). In a study conducted by Gogoi and Prusty (2013), the symptom of abnormal vaginal discharge was recorded as 9.7 percent among the tribal population in Assam. From their study, the prevalence of vaginal discharge was reported to be high than any other gynaecological morbidities among all the studied tribal and non-tribal populations in the North-eastern region of India. Morang (2019) reported the problem of abnormal discharge among 14.42 percent of Mishing mothers of Assam.

Out of all the socio-demographic characters considered, the chi-square test found out statistically significant differences with regards to the prevalence of abnormal vaginal discharge according to the present age group of the respondents. The prevalence of abnormal vaginal discharge was recorded to be higher among middle-aged mothers. Among the young mothers also the occurrence was found to be common. The study of Patel *et al.,* (2005) and Guntoory *et al.,* (2017) also found a higher percentage of vaginal discharge among younger mothers. The occurrence of abnormal vaginal discharge was quite common in the rural area as compared to the urban area. One of the reasons may be the rural respondents used to go for community fishing in the nearby ponds, marshy lands, etc. for which hygienic condition is not possible to maintain. The similarity of the present finding is observed with the study of Hazarika (2019) among the Kaibartas of Assam. A higher prevalence of abnormal discharge among less-educated respondents was reported from the present study. This finding was consonant with the study of Sadhu *et al.,* (2001), Chaudhary *et al.,* (2012), and Guntoory *et al.,* (2017). However, homemakers suffered higher proportions of abnormal discharge than working mothers. Dibongiya mothers with higher numbers of parity had reported less percentage of vaginal discharge. It was in contrast with the study of Sadhu *et al.,* (2001) among the women in weaker sections in Kolkata Metropolitan areas. More number of mothers from nuclear families had reported the complaint of vaginal discharge than the urban setup. The proportion of the sufferers was found to be more among the mothers who had adopted the process of sterilization as a family planning measure. The study of Pant *et al.,* (2008), Chaudhary *et al.,* (2012), and Guntoory *et al.,* (2017) in various parts of India also reported the same finding. Experience of abnormal discharge was more common among the Dibongiya mothers who used both cloth and disposable material during menstruation. Amongst them, most of the mothers used an old piece of cloth and had the occasional use of disposable napkins during periods.

Dibongiya Deori mothers were also experiencing some other symptoms associated with leucorrhea. Out of the total respondents, 17.86 percent of them had suffered from at least one symptom. However, some of them had the experience of more than one symptom. Feeling of weakness and pain in the lower abdomen without menses were the common symptoms associated with abnormal discharge. The study of Sadhu *et al.,* (2001) recorded the problems of itching and/or burning sensation in the genital region followed by backache, lower abdominal pain, bad odour, and fever accompanied with vaginal discharge. The study of Patel *et al.,* (2005) and Chaudhary *et al.,* (2012) also found a strong association of vaginal discharge with some other gynaecological symptoms like itching and blisters in the genital area, pain in the lower abdomen, and burning during micturition.

The accompanying problems of vaginal discharge were quite common among the respondents with less educational attainment, homemakers, sterilized, and belonged to nuclear families. From the present study, a clear-cut picture could not be drawn in regards to the prevalence of abnormal discharge with several parities. On the contrary, the study of Sadhu *et al.,*

(2001) found a positive association between the number of parity and prevalence of various symptoms along with vaginal discharge. The study of Kulkarni *et al.,* (2005) and Chaudhary *et al.,* (2012) showed a significant association of discharge with high parity. But the study conducted in Goa by Patel *et al.,* (2005) reported a contrast finding in this regard.

More numbers of respondents from the rural area had suffered from at least one variety of symptoms along with vaginal discharge. However, the accompanying problems of vaginal discharge were found to be common among those who used disposable sanitary napkins during periods. The frequency of changing their napkins during periods may be one of the reasons behind this.

Conclusion

The self-concern in women for their own health needs is of utmost importance. Many background characters play an important role in shaping the health of women in general and reproductive health in particular. The present study is also a witness of the role of various socio-demographic factors on the problem of abnormal vaginal discharge among the Dibongiya Deori mothers of Assam. Creating community awareness about health care facilities will be helpful to overcome such kind of problems. As the number of studies in North-east India is found to be limited in this regard, the present study may be a base for the policymakers for the betterment of society.

Acknowledgement

Gratitude goes to all the respondents of Lakhimpur district of Assam, with their active participation only we could complete the data collection smoothly and timely.

References

Bhalla, P, Chawla, R, Garg, S, Singh, MM, Raina, U, Bhalla, R & Rodhanit, P 2007, 'Prevalence of bacterial vaginosis among women in Delhi, India', *Indian Journal of Medical Research*, vol. 125, no. 2, pp. 167-172.

Bhatia, JC, Cleland, J, Bhagavan, L & Rao, NS 1997, 'Levels and determinants of gynecological morbidity in a district of South India', Studies in Family Planning, vol. 28, pp. 95-103.

Chaudhary, V, Kumar, R, Agrawal, VK, Singh, A, Narula, R & Sharma, M 2012, 'Prevalence and determinants of vaginal discharge among women of reproductive age group in Tertiary Care Hospital of Northern India', *National Journal of Community Medicine*, vol. 3, no. 4, pp. 661-665.

Dan, M, Kaneti, N, Levin, D, Poch, F & Samra, Z 2003, 'Vaginitis in a gynecologic practice in Israel: causes and risk factors', *The Israel Medical Association Journal*, vol. 5, no. 9, pp. 629-632.

Department of Welfare of Plains Tribe and Backward Classes 2007. Govt. of Assam.

Gittlesohn, J, Bentley, ME & Pelto, PJ (eds.) 1994, *Listening to women talk about their health: issues and evidence from India*, Har-Anand Publications, New Delhi, pp. 21-36.

Guntoory, I, Tamaraba, NR, Nambaru, LR & Kalavakuri, AS 2017, 'Prevalence and socio-demographic correlates of vaginal discharge among married women of reproductive age group at a teaching hospital', *International Journal of Reproduction, Contraception, Obstetrics and Gynecology*, vol. 6, no. 48, pp. 4840-4846.

Hazarika, J 2019, *Reproductive health and nutritional status of Kaibarta women of Assam an anthropological approach*. Ph. D. thesis, Department of Anthropology, Dibrugarh University.

Joshi, A, Dhapola, M & Pelto, PJ 2008, 'Gynaecological problems: perceptions and treatment-seeking behaviours of rural Gujarati women' in *Reproductive health in India new evidence*, eds. MA Koenig, S Jejeebhoy, JC Cleland & B Ganatra, Rawat Publications, Jaipur, pp. 133-158.

Kanani, S, Latha, K & Shah, M 1994, 'Application of qualitative methodologies to investigate perceptions of women and health practitioners regarding women's health disorders in Baroda slums' in *Listening to women talk about their health: issues and evidence from India*, eds. J Gittelsohn, ME Bentley & PJ Pelto, Har-Anand Publications, New Delhi, India.

Kaur, J & Kapoor, AK 2014, 'Perceptions and knowledge about leukorrhea in a slum dwelling South Asian Community', *Journal of Family and Reproductive Health*, vol. 8, no. 1, pp. 45–52.

Kulkarni, RN & Durge, PM 2005, 'A study of leucorrhea in reproductive age group women of Nagpur city', *Indian Journal of Public Health*, vol. 49, no. 4, pp. 238-239.

Lamont, RF, Morgan, DJ, Wilden, SD & Taylor-Robinson, D 2000, 'Prevalence of bacterial vaginosis in women attending one of three general practices for routine cervical cytology', *International Journal of STD & AIDS*, vol. 11, no. 8, pp. 495-498.

Loretta, B 1995, 'Reproductive tract infections and abortion among adolescent girls in rural Nigeria', *Lancet*, vol. 345, pp. 300-304.

Morang, P 2019, *A study on nutritional health and nutritional status of the Mishing women of Assam*. Ph. D. thesis, department of anthropology, Dibrugarh University.

Oomman, N 1996, *Poverty and pathology: comparing rural Rajasthani women's ethnomedical models with biomedical models of reproductive morbidity: implications for women's health in India*. Dissertation, John Hopkins University, Baltimore, USA.

Pant, B, Singh, JV, Bhatnagar, M, Garg, SK, Chopra, H & Bajpai, SK 2008, 'Social correlates in reproductive tract infections among married women in rural area of Meerut', *Indian Journal of Community Medicine*, vol. 33, no. 1, pp. 52-53.

Patel, T 1998, *Fertility behaviour: population and society in a Rajasthan village*, Oxford University Press, New Delhi, India.

Patel, V, Pednekar, S, Weiss, H, Rodeigues, M, Barros, P, Nayak, B, Tanksale, V, West, B, Nevrekar, P, Kirkwood, BR & Mabey, D 2005, 'Why do women complain of vaginal discharge? A population survey of infectious and psychological risk factors in a South Asian community', *International Journal of Epidemiology*, vol. 34, pp. 853-862.

Rakoff, AE & Bland, PB 1940, 'Leukorrhea: clinical and therapeutic aspect', *The Journal of the American Medical Association*, vol. 115, no. 12, pp. 1013-1018.

Sadhu, S, Bhattacharya, BN, Majumder, PK & Mukherjee, KL 2001, 'Reproductive morbidity among women in weaker sections in Kolkata metropolitan area', *Demography India*, vol. 30, no. 2, pp. 167-191.

Wasserheit, JN, Harris, JR, Chakraborty, J, Kay, BA & Mason, KJ 1989, 'Reproductive tract infections in a family planning population in rural Bangladesh', *Studies in Family Planning*, vol. 20, no. 2, pp. 69-80.

Chapter - 6

Assessment of Health Status of Female Tea Garden Workers in Dibrugarh District of Assam

Bhaskar Das, Rumi Deori, and Anju Barhai Teli

The background

The future of a nation depends on the health of its citizens and thus, very rightly it is one of the major components to measure the Human Development Index (HDI). The World Health Organization (WHO) in 1948 defines health in a broader sense as "a state of complete physical, mental and social well-being and not merely the absence of disease or infirmity". Although this definition faced criticism at some quarters as being vague and for use of the word "complete" nonetheless, this remains the accepted definition worldwide. What is said about the states of India is equally true for Assam as well. With a population of more than 31 million (Census, 2011) Assam is well represented by multi-ethnic, multi-linguistic, and multi-religious groups. This includes the tea garden workers which constitute more than 1/5th of the state's total population. In India, tea was discovered in Assam in the early 19th century and thanks to these tea garden workers, Assam produces around 55.0 percent of India's total production of tea (Sengupta, 2009). The workers are the descendants of both tribal and caste groups who were brought to Assam from present-day Jharkhand, Odisha, West Bengal, Telangana, and Chhatisgarh by the British colonial planters to work in the tea garden industry as cheap indentured labourers. Although tea is the major agro-industry in Assam which contributes immensely to the state's economy unfortunately the condition of the tea workers since their migration to Assam is not less than pathetic. Poor socio-economic conditions, ignorance due to illiteracy, overcrowded and unhygienic living conditions in the residential colonies (earlier referred to as *Coolie* lines) make them vulnerable to various communicable diseases and malnutrition. Several studies have highlighted that undernutrition, anaemia, and hypertension among other diseases are high among the tea garden workers (Biswas *et al.,* 2002; Hazarika *et al.,* 2002; Medhi *et al.,* 2006; Das and Sengupta, 2009; Das *et al.,* 2012; Timung and Sarmah 2013; Panyang *et al.,* 2018; Times of India, 2018; Konwar *et al.,* 2019). Even the Assam Human Development Report (2003) specifically acknowledged the fact that the health status of the tea workers is much below the state average. The state has around 800 tea estates and they are mostly in the upper Assam region with the Dibrugarh district having the highest proportion. Studies show that like the rest of Assam, malnutrition, anaemia, and hypertension are widespread among the tea garden workers in Dibrugarh district (Hazarika *et al.,* 2002; Medhi *et al.,* 2006; Das and Sengupta, 2009; Das *et al.,* 2012; Panyang *et al.,* 2018).

Anthropometric measurements and indices are reported to be good indicators for malnutrition (James *et al.,* 1994; Biswas *et al.,* 2002; Chakraborty, Bose and Koziel 2011; Das and Bose 2012; Borah and Sengupta, 2019) but whether haemoglobin level and other haematological parameters are in good agreement with the former is not known (Das *et al.,* 2012). The present study, therefore, tries to assess the health status of female tea garden workers in the Dibrugarh district through anthropometric, haemoglobin, and a physiological parameter namely arterial blood pressure.

Objectives of the study

The study was carried out with the following objectives in mind:

(i) To assess the nutritional status of the female tea garden workers through anthropometric parameters namely body mass index (BMI) and mid-upper arm circumference (MUAC);

(ii) to examine the haematological parameters of the subjects;

(iii) to measure the arterial blood pressure of the subjects; and

(iv) to evaluate whether BMI and MUAC categories have a significant difference with the respective mean haematological parameters and arterial blood pressure.

Subjects and methods

At the outset, the tea gardens were surveyed and the purpose of the study and its importance was explained to the tea garden authorities. After obtaining consent, the tea workers' colonies were surveyed and the purpose of the study and its importance was explained to the workers. Female tea garden workers were considered under the present study and care was taken to include only non-pregnant women of reproductive age. Prior informed consent was taken from the subjects and their participation was voluntary.

Altogether 234 female tea garden workers from Bahubali (n=101), Mohini (n=95), and Jamuna (n=38) Tea Estates of Dibrugarh district were studied and the findings were analyzed keeping the above aim and objectives in mind. It may be noted that name of the Tea Estates is changed in this paper due to requests from the tea garden authorities. The tea estates are approximately 45 km, 30 km, and 20 km distance from Dibrugarh town. The workers of the plantations are heterogeneous consisting of both tribal and caste populations migrated from states like Odisha, Bihar, West Bengal, Madhya Pradesh, etc. The selection of subjects was done by random sampling from each of the tea gardens.

Anthropometric measurements namely height, weight, and MUAC were taken from the subjects (with anthropometer, weighing machine, and tape respectively) followed by a collection of blood samples (2 ml approximately) by vein puncture. The blood sample was collected by a lab technician under the supervision of a medical practitioner from the Assam Medical College and Hospital, Dibrugarh. A general medical examination including measurement of arterial blood pressure (through standard mercury sphygmomanometer) was carried out on the subjects before collecting the blood sample. Data were collected by organizing camps at the respective tea gardens. Anthropometric assessment of nutritional status was done through BMI (weight in kilogram divided by height in meter^2) and MUAC. Analyses (haematological parameters) of the blood samples were carried out through Cell Counter (Sysmex) at the Biochemistry Department of the Assam Medical College and Hospital, Dibrugarh. Data were analyzed with the help of statistical software SPSS 17.0.

Findings

BMI is classified as per WHO's Asian criteria. It is found that 58.1 percent of the female workers are underweight while 34.6 percent have a normal BMI (Table 6.1). Classification of MUAC is done as per James *et al.*, (1994). The study revealed that 34.6 percent of workers have MUAC <22.0 and thus, are considered undernourished. Haemoglobin concentration of the female workers is classified as per the World Health Organization (WHO) classification of anaemia (Table 6.3). Anaemia is categorized as mild, moderate, and severe. The study shows that 79.92 percent of the female workers are anaemic of which 32.05 percent are mild, 42.31 percent

Table 6.1: Body Mass Index of the female tea garden workers

Nutritional status	*Asian criteria BMI (in kg)*	*N*	*Percentage*	*Mean*	*SD*
Underweight	<18.5	136	58.1	16.59	1.29
Normal	18.5 – 22.9	81	34.6	19.86	1.07
Overweight	23.0 – 24.9	6	2.6	23.78	0.54
Obese I/ Pre-obese	25.0 – 29.9	11	4.7	27.1	1.69
Total		**234**	**100.0**		

Table 6.2: Mid Upper Arm Circumference of the female tea garden workers

Nutritional status	*MUAC cut-off (in cm)*	*N*	*Percent*	*Mean*	*SD*
Undernourished	<22.0	81	34.6	20.65	1.06
Normal	22.0 & above	153	65.4	24.48	2.24
Total		**234**	**100.0**		

are moderate and 5.56 percent are severely anaemic. Among the mildly anaemic workers, the mean haemoglobin concentration is 11.42, and their mean corpuscular volume (MCV) and mean corpuscular haemoglobin (MCH) points towards microcytic hypochromic anaemia although red blood cells (RBC), mean corpuscular haemoglobin concentration (MCHC), and red cell distribution width-coefficient of variation (RDW-CV) are within the normal range. Among the moderately and severely anaemic workers, the mean haemoglobin is 10.21 and 6.56 respectively.

The other haematological parameters of both these groups are indicative of microcytic hypochromic anaemia with elevated RDW-CV. They also have low mean RBC. It is found that only 20.08 percent of the workers are non-anemic and their mean haemoglobin concentration of 12.78. Their other haematological parameters like RBC, MCV, MCH, MCHC, and RDW-CV are also within the normal range. It is found that 20.1 percent have elevated systolic blood pressure whereas stage 1 and stage 2 high systolic blood pressure accounts for 20.6 percent of the workers (Table 6.4). On the other hand, 49.5 percent of the workers are found to have stage 1 and stage 2 high diastolic blood pressure (Table 6.5).

The BMI categories are subjected to the Kruskal Wallis test to evaluate whether significant differences exist between the respective mean haematological parameters. The P values however, do not show any significant differences in the haematological parameters (namely

Table 6.3: Haematological parameters of the female tea garden workers

Range of haemoglobin concentration	*Hb (g/dl)*		*RBC ($10^6/\mu L$)*		*MCV (fl)*		*MCH (pg)*		*MCHC (g/dl)*		*RDW-CV (%)*	
	Mean	*SD*	*Mean*	*SD*	*Mean*	*SD*	*Mean*	*SD*	*Mean*	*SD*	*Mean*	*SD*
Cut-off/ Non-anaemia 12.0 g/dl (n = 47)	12.78	0.81	4.65	0.53	81.69	11.93	27.9	3.07	34.22	5.27	14.18	1.81
Mild anaemia 11.0 – 11.9 g/dl (n = 75)	11.42	0.29	4.52	0.59	77.17	10.68	25.76	3.67	33.1	1.44	14.42	1.41
Moderate anaemia 8.0 – 10.9 g/dl (n = 99)	10.21	0.62	4.32	0.53	73.81	10.32	24.73	6.69	32.31	1.38	16.35	2.73
Severe anaemia < 8.0 g/dl (n = 13)	6.56	1.45	3.61	0.58	62.98	7.54	18.2	3.63	28.68	3.01	23.61	2.81

Table 6.4: Systolic blood pressure of the female tea garden workers

Systolic Blood Pressure	*Range (mm Hg)*	*N*	*Percent*	*Mean*	*SD*
Normal	80-119	139	59.4	104.89	7.06
Elevated blood pressure	120-129	47	20.1	120.34	1.46
Stage 1 High blood pressure	130-139	24	10.3	130.0	0
Stage 2 High blood pressure	140-180	24	10.3	148.08	11.85
Total		**234**	**100.0**		

Table 6.5: Diastolic blood pressure of the female tea garden workers

Diastolic Blood Pressure	*Range (mm Hg)*	*N*	*Percent*	*Mean*	*SD*
Normal	60-79	118	50.4	69.58	4.01
Stage 1 High blood pressure	80-89	78	33.3	80.82	2.02
Stage 2 High blood pressure	90-120	38	16.2	92.68	4.29
Total		**234**	**100.0**		

Table 6.6: Kruskal Wallis test: BMI categories with haematological parameters

BMI Categories	*N*	*Haemoglobin*			
		Mean	*SD*	*Chi-Square value*	*P-value*
Underweight	136	10.9	1.54	4.3	0.231
Normal	81	10.96	1.72		
Overweight	6	11.38	1.27		
Obese I/ Pre-obese	11	10.34	0.89		
BMI Categories	*N*	*RBC*			
		Mean	*SD*	*Chi-Square value*	*P-value*
Underweight	136	4.4	0.61	0.88	0.829
Normal	81	4.42	0.55		
Overweight	6	4.65	0.64		
Obese I/ Pre-obese	11	4.39	0.76		
BMI Categories	*N*	*MCV*			
		Mean	*SD*	*Chi-Square value*	*P-value*
Underweight	136	75.9	11.5	0.34	0.953
Normal	81	75.94	11.77		
Overweight	6	74.15	9.07		
Obese I/ Pre-obese	11	75.95	10.1		

(Cond...)

BMI Categories	*N*	*MCH*			
		Mean	*SD*	*Chi-Square value*	*P-value*
Underweight	136	25.21	3.89	1.3	0.729
Normal	81	25.78	7.69		
Overweight	6	24.13	2.94		
Obese I/ Pre-obese	11	24.23	3.77		
BMI Categories	*N*	*MCHC*			
		Mean	*SD*	*Chi-Square value*	*P-value*
Underweight	136	32.9	3.61	3.29	0.35
Normal	81	32.63	1.85		
Overweight	6	32.52	0.47		
Obese I/ Pre-obese	11	31.84	1.61		
BMI Categories	*N*	*RDW-CV*			
		Mean	*SD*	*Chi-Square value*	*P-value*
Underweight	136	15.6	3.08	7.38	0.061
Normal	81	15.71	3.28		
Overweight	6	15.58	1.76		
Obese I/ Pre-obese	11	16.97	1.84		

Table 6.7: Kruskal Wallis test: BMI categories with systolic and diastolic blood pressure

BMI Categories	*N*	*Systolic Blood Pressure*			
		Mean	*SD*	*Chi-Square value*	*P-value*
Underweight	136	113.22	15.12	15.03	0.002*
Normal	81	116.37	16.05		
Overweight	6	109.0	11.51		
Obese I/ Pre-obese	11	130.18	14.01		
BMI Categories	*N*	*Diastolic Blood Pressure*			
		Mean	*SD*	*Chi-Square value*	*P-value*
Underweight	136	75.66	9.28	16.04	0.001*
Normal	81	78.32	8.86		
Overweight	6	76.33	4.97		
Obese I/ Pre-obese	11	85.82	7.67		

*significant at 0.05 level of probability

haemoglobin concentration, RBC, MCV, MCH, MCHC, and RDW-CV) about the different BMI categories (Table 6.6). However, both the P values are found to be significant when the mean systolic and diastolic blood pressures of the workers under different BMI categories are compared (Table 6.7). Similarly, the mean haematological parameters against the two MUAC categories are subjected to the Kruskal Wallis test. It is found that except for RBC (P = 0.004), no other P values show any significant difference in the rest of the haematological parameters (Table 6.8). Likewise, when the mean systolic and diastolic blood pressures are subjected to analysis of variance tests about the two MUAC categories, it is found that both the P values are significant (Table 6.9).

Table 6.8: Kruskal Wallis test: MUAC categories with haematological parameters

MUAC Categories	*N*	*Haemoglobin*			
		Mean	***SD***	***Chi-Square value***	***P-value***
Undernourished	81	10.70	1.53	2.13	0.145
Normal	153	11.02	1.59		
MUAC Categories	***N***	***RBC***			
		Mean	***SD***	***Chi-Square value***	***P-value***
Undernourished	81	4.24	0.60	8.37	0.004*
Normal	153	4.50	0.57		
MUAC Categories	***N***	***MCV***			
		Mean	***SD***	***Chi-Square value***	***P-value***
Undernourished	81	76.84	11.31	0.88	0.349
Normal	153	75.35	11.48		
MUAC Categories	***N***	***MCH***			
		Mean	***SD***	***Chi-Square value***	***P-value***
Undernourished	81	25.59	4.08	2.50	0.114
Normal	153	25.20	6.10		
MUAC Categories	***N***	***MCHC***			
		Mean	***SD***	***Chi-Square value***	***P-value***
Undernourished	81	32.81	2.03	2.36	0.124
Normal	153	32.71	3.38		
MUAC Categories	***N***	***RDW-CV***			
		Mean	***SD***	***Chi-Square value***	***P-value***
Undernourished	81	15.92	3.43	0.05	0.821
Normal	153	15.58	2.89		

*significant at 0.05 level of probability

Table 6.9: Kruskal Wallis test: MUAC categories with systolic and diastolic blood pressure

MUAC Categories	*N*	*Systolic Blood Pressure*			
		Mean	*SD*	*Chi-Square value*	*P-value*
Undernourished	81	112.15	15.87	4.54	0.033*
Normal	153	116.51	15.45		
MUAC Categories	*N*	*Diastolic Blood Pressure*			
		Mean	*SD*	*Chi-Square value*	*P-value*
Undernourished	81	74.25	9.32	14.56	0.00*
Normal	153	78.58	8.85		

Discussion

BMI is a widely used indicator to determine nutritional status in adults. MUAC is another anthropometric measure that is used to evaluate adult nutritional status. Haemoglobin concentration is undoubtedly an excellent indicator to assess the general health status of an individual. In the present study, 58.1 percent of the workers are found to have a low BMI (underweight). Although the prevalence is relatively less than those reported by Medhi *et al.,* (2006) and Das *et al.,* (2012), the situation is still considered as critical as per WHO (1995). Classification of MUAC shows that 34.6 percent of workers are undernourished. About haemoglobin concentration, 79.92 percent of the female workers are found to be anaemic, of which 32.05 percent have mild anaemia and the rest suffer from moderate (42.31%) and severe (5.56%) anaemia. The overall prevalence of anaemia corroborates with earlier studies carried out in the district (Medhi *et al.,* 2006, Das *et al.,* 2012, Panyang *et al.,* 2018) however they recorded a higher frequency of severe anaemia in the population. According to WHO (2008), a cut-off value of $\geq$ 40.0 percent prevalence of anaemic in a population poses a severe public health problem.

Low BMI is associated with a markedly increased risk for ill health, poor physical performance, lethargy, and death in extreme circumstances (NLIS, 2010). In women, anaemia is associated with morbidity and increased risks for reproductive wastage, maternal and child mortality. While there may be several causes for anaemia but it is found that iron deficiency is the commonest of all the anaemias. Earlier studies have shown that nutritional (including iron, folate, and vitamin B^{12} deficiency) anaemia, helminthiasis, and haemglobinopathies induced anaemia is frequently found among the tea garden workers of Assam (Medhi *et al.,* 2006, Das *et al.,* 2012, Panyang *et al.,* 2018). Test of significance (Kruskal Wallis) shows that there is no significant difference between the BMI categories and the respective mean haematological parameters. Similarly about the MUAC categories also, the P values (except RBC, P = 0.004) do not show any significant difference. However, arterial blood pressure (systolic and diastolic) shows a significant difference between both BMI and MUAC. Our findings suggest that arterial blood pressure depends both on the BMI and MUAC but in the case of haemoglobin concentration it is not found to be so. Among the tea garden workers though, arterial blood pressure may not always depend on BMI or MUAC as the prevalence of hypertension is frequent in them primarily due to intake of salted tea, extra salt in meal, and consumption of locally prepared alcohol (Hazarika *et al.,* 2002; Medhi *et al.,* 2006; Das and Sengupta, 2009). The present study also found 20.6 percent of the workers with high systolic blood pressure (stage 1 and 2) and 49.5 percent with high diastolic blood pressure (stage 1 and 2).

Conclusion

For the fact that Assam still has the highest maternal mortality rate and second-highest rate in terms of infant mortality (MMR, 2015-17; SRS, 2015-17), a multifaceted approach will be helpful to assess the health status of a population at the micro-level. The present study has shown that haemoglobin concentration is independent of both BMI and MUAC whereas in the case of arterial blood pressure it is not found to be so. However, more studies using additional anthropometric indices/ measurements and a larger sample are required to come to a firm conclusion.

Acknowledgement

The authors express their gratitude to the UGC-NERO, Guwahati for providings funds for the project

References

Biswas, D, Hazarika, NC, Doloi, P & Mahanta, J 2002, 'Study on the nutritional status of tea garden workers of Assam with special emphasis on body mass index (BMI) and central obesity', *Journal of Human Ecology,* vol. 13, pp.299-302.

Borah J & Sengupta, S 2019, Nutritional status among the adult Dibongiya Deori women of Lakhimpur district, Assam. In: *Growth and Nutritional Studies: Perspective from North-east India*, Sengupta S, Dutta D (Eds), Gyan Publishing House, New Delhi.

Chakraborty, R, Bose, K & Koziel, S 2011, 'Use of mid-upper arm circumference in determining undernutrition and illness in rural adult Oraon men of Gumla District, Jharkhand, India', *Rural and Remote Health,* vol.11, pp.1754.

Das, B, Sengupta, B, Chakrabarti S, Rudra, T & Sengupta, S 2012, 'Incidence of anaemia among the female tea garden workers in a tea plantation in upper Assam, India', *Journal of Indian Medical Association,* vol.110, pp. 84-87.

Das, B & Sengupta, S 2009, Health of tea garden workers: A case study in Upper Assam. In: *Tribes of North-East India: Issues and Challenges*, Medhi BK, Athparia RP, Jose K (Eds), Omsons Publications, New Delhi, pp.145-153.

Das, S & Bose, K 2012, 'Nutritional assessment by mid-upper arm circumference of Santal adults of Purulia, West Bengal, India', *Collegium Antropologicum,* vol.36, pp.581-584.

Hazarika, NC, Biswas, D, Narain, K, Kalita, HC & Mahanta, J 2002, 'Hypertension and its risk factors in tea garden workers of Assam', *National Medical Journal of India,* vol.15, pp.63-68.

James, WP, Mascie-Taylor, GC, Norgan, NG, Bistrian, BR, Shetty, PS & Ferro-Luzzi, A 1994, 'The value of arm circumference measurements in assessing chronic energy deficiency in Third World adults', *European Journal of Clinical Nutrition,* vol.48, pp.883-894.

Konwar, P, Vyas, N, Hossain, SS, Gore, MN & Choudhury, M 2019, 'Nutritional status of adolescent girls belonging to the tea garden estates of Sivasagar district, Assam, India', *Indian Journal of Community Medicine,* vol.44, pp.238-242.

Medhi, GK, Hazarika, NC, Shah, B & Mahanta, J 2006, 'Study of health problems and nutritional status of tea garden population of Assam', *Indian Journal of Medical Science,* vol.60, pp.496-505.

NLIS 2010, *Nutrition Landscape Information System (NLIS) country profile indicators: interpretation guide.* WHO.

Panyang, R, Teli, AB & Saikia, SP 2018, 'Prevalence of anemia among the women of childbearing age belonging to the tea garden community of Assam, India: A community-based study', *Journal of Family Medicine and Primary Care,* vol.7, pp.734-738.

Sengupta, S 2009, *The Tea Labourers of North-east India*. Mittal Publications, New Delhi.

Times of India 2018, Anaemia high among women in Assam's tea gardens: Study. Guwahati, July 24.

Timung, J & Sarmah, J 2013, 'Nutrition, health, and hygienic practice of women tea plantation workers of Assam', *Asian Journal of Home Science,* vol.8, pp.421-424.

WHO 1995, *Physical status: the use and interpretation of anthropometry.* Report of a WHO Expert Committee. Technical Report Series No. 854.

WHO 2008, *Worldwide prevalence of anaemia 1993-2005 WHO Global Database on Anaemia.*

Chapter - 7

Hypertension and its Behavioural Risk Factors among Tea Garden Workers of Assam

Anand Sharma and Prasanta Kumar Borah

Introduction

Hypertension is an important risk factor for cardiovascular disease and is a major public health problem in developing countries around the world. As reported by the World Health Organization, hypertension (HTN) is the third 'killer' disease, accounting for one in every eight deaths worldwide (WHO, 2002). It has been expected that the number of hypertensive may rise from 118 million in 2000 to 214 million in 2025 (Kearney *et al.,* 2005).

Rapid unplanned urbanization also tends to promote the development of hypertension as a result of unhealthy environments that encourage the consumption of fast food, sedentary behaviour, tobacco use, and the harmful use of alcohol. Finally, the risk of hypertension increases with age due to the stiffening of blood vessels, although ageing of blood vessels can be slowed through healthy living, including healthy eating and reducing the salt intake in the diet (WHO, 2013).

Health-risk behaviors, which contribute to leading causes of mortality and morbidity, often are established during youth, extend into adulthood, are interrelated, and preven Table (Kann *et al.,* 2000). Globally, tobacco and alcohol rank 4th and 5th, respectively, as risk factors for avoidable disease measured by disability-adjusted life years (DALY; WHO, 2002).

Tea is one of the largest agricultural industries in India. There are reports of a rapid rise in tobacco and alcohol-related disease in people in the tea industry in Assam, north-eastern India. However, available records indicate a higher prevalence of hypertension among this population in comparison to other studies (Hazarika *et al.,* 2002).

The tea garden workers (TGW) community of Assam is a migratory population of North-east India engaged in the tea garden industry. Earlier studies revealed a very high prevalence of hypertension (60.8% in the age group $\geq$ 30 years; Hazarika *et al.,* 2002). Hospital-based data also reveals a high prevalence of stroke from this community (personal communication with a clinician). The majority (>85.0%) of this population consumes extra salt (salt as a side dish) in a reasonably high amount (approx. 15-20 gm/day) and it was found to be a significant risk factor for hypertension in the earlier studies. The present study was conducted to find out the habit of substance use and the risk of hypertension among tea garden workers of Assam.

Materials and methods

A community-based cross-sectional study in a tea garden of Dibrugarh district, Assam by using a multistage stratified sampling method was conducted. A total of 901 adults in the age group of 18 years and above were collected for the study. Information regarding socio-demographic variables including age, marital status, religion, education, occupation, type of work engaged, smoking habit, the habit of taking extra salt, and alcohol consumption were recorded by personal interview. Height and weight were recorded by a trained person using standard methodologies. Anthropometric rod and platform balance (SECA) were used to measure height and weight respectively. Body mass index (BMI) was calculated as weight in kg divided by height in meter squared.

Blood pressure was measured using electronic blood pressure equipment (Omron model HEM 907) following standard guidelines. Three readings at an interval of 5-10 minutes for each participant and the average of three readings were used for analysis. Participants who smoked and consumed alcohol were allowed to take rest for about 30 minutes before the recording of BP. Hypertension was defined as per JNC VII guidelines.

Inclusion criteria

The following inclusion criteria were adopted for the recruitment of the study subjects:

- Tea garden worker of either gender and aged > = 18 years
- Subjects proving informed consent.

Exclusion criteria

- Subjects unable to provide informed consent
- Subjects <18 years
- Pregnant women
- Subjects with a history of stroke, transient ischemic attack, congestive cardiac failure, myocardial infarction, and angina.
- Patients with known diseases of kidney, liver, and malignancy.

Results

Table 7.1: Socio-demographic characteristics of tea garden workers of Assam according to blood pressure status

Variable	*Subgroups*	*Group*		*p-value*
		Hypertensive N (%)	*Non-hypertensive N(%)*	
Sex	Male	290(62.0)	228(52.7)	0.05
	Female	178(38.0)	205(47.3)	
	Total	468(51.9)	433(48.1)	
Age group	18-30	100(21.4)	234(54.0)	0.000
	31-40	205(43.8)	117(27.0)	
	41-50	97(20.7)	56(12.9)	
	51 and above	66(14.1)	26(6.0)	
	Total	468(51.9)	433(48.1)	
Educational status	Non-literate	320(69.1)	248(58.1)	0.007
	Primary	121(26.1)	153(35.8)	
	Higher Secondary	19(4.1)	24(5.6)	
	Graduate	3(0.6)	1(0.2)	
	Professional	0 (0.0)	1(0.2)	
Income	Upto 2000	193(41.2)	228(52.7)	0.005
	2000-3000	72(15.4)	53(12.2)	
	3000-4000	16(3.4)	20(4.6)	
	4000-5000	4(0.9)	7(1.6)	
	5000+	3(0.6)	3(0.7)	
	Dependant	180(38.5)	122(28.2)	
Occupation	Cultivation	6(1.3)	8(1.8)	0.000
	Service	247(52.8)	264(61.0)	
	Housewife	42(9.0)	61(14.1)	
	Daily wages	25(5.3)	30(6.9)	
	Others	148(31.6)	70(16.2)	
Marital status	Married	396(84.6)	345(79.7)	0.000
	Unmarried	32(6.8)	68(15.7)	
	Divorced	1(0.2)	7(1.6)	
	Separated	0 (0.0)	1(1.2)	
	Widow/Widower	39(8.3)	12(2.8)	

Table 7.1 shows that the prevalence of hypertension was higher in males (62.0%) than in females (38.0%). The difference in the prevalence of hypertension in both genders was statistically significant (<0.05). The overall prevalence of hypertension was 51.9 percent and the prevalence of hypertension was found to be highest in the age group of 31-40 years (<0.001). Educational status states that 69.1 percent of hypertensive are non-literates followed by primary (26.1%), higher secondary (4.1%). 41.2 percent of hypertensive have an income of up to Rs. 2000 while 52.7 percent of non-hypertensive (<0.005). The majority of the prevalence of hypertension was found in service holders (52.8%) followed by others (31.6%), housewives (9.0%), daily wages (5.3%), and cultivation (1.3%) (<0.001).

Table 7.2: Distribution of clinical characteristics of tea garden workers according to blood pressure status

Variables		*Hypertensive (%)*	*Non-hypertensive (%)*	*Total (%)*	*p-value*
Body mass index	< 18.5	194(41.5%)	185(42.8%)	379(42.1%)	0.680
	18.5-24.9	259(55.3%)	233(53.9%)	492(54.7%)	
	25-29.9	14(3.0%)	11(2.5%)	25(2.8%)	
	≥ 30	1(0.2%)	3(0.7%)	4(0.4%)	
Tobacco consumption, yes	Yes	349(74.6)	228(52.7)	577(64.0)	0.000
	No	119(25.4)	205(47.3)	324(36.0)	
Alcohol consumption	Yes	296(63.2)	198(45.7)	494(54.8)	0.000
	No	172(36.8)	235(54.3)	407(45.2)	
Smoking consumption	Yes	58(12.4)	28(6.5)	86(9.5)	0.002
	No	410(87.6)	405(93.5)	815(90.5)	
Extra salt	Yes	340(72.6)	284(65.5)	624(69.3)	0.022
	No	128(27.4)	149(34.4)	277(30.7)	
Physical activity	Light	43(9.2)	62(14.3)	105(11.7)	0.030
	Moderate	381(81.4)	323(74.6)	704(78.1)	
	Heavy	44(9.4)	48(11.1)	92(10.2)	

Table 7.2 states that the distribution of tea garden workers according to Body Mass Index shows that 55.3 percent of hypertensive were found to be normally followed by underweight (41.5%), overweight (3.0%), and obese (0.2%). Almost 75.0 percent of the tea garden community having a habit of tobacco chewing was having hypertension74.6 percent while 25.4 percent not having a habit of tobacco chewing was having hypertension (<0.001). Among 494 alcoholic users, 63.2 percent were hypertensive and 45.1 percent were non-hypertensive (<0.001). The prevalence of hypertension among smokers was 12.4 percent while 87.6 percent were non-smokers (<0.005). 72.6 percent of hypertensive patients were taking

extra salt intake while 27.4 percent of hypertensive patients were not taking extra salt (<0.005). A higher prevalence of hypertension was found with moderate activity (81.4%) followed by heavy activity (9.4%) and light activity (9.2%) (<0.005).

Table 7.3: Distribution of anthropometric indicators in tea garden workers according to blood pressure status

Variables	*Normotensive(N=433) (Mean ± S.D)*	*Hypertensive(N=468) (Mean ± S.D)*	*P-value*
SBP(mm/Hg)	121.8±9.3	156±19.0	0.000
DBP(mm/Hg)	74.8±8.1	91.7±11.5	0.000
Age(years)	33.7±16.7	46.0±15.5	0.000
Height(cm)	154.5±8.2	154.3±8.4	0.682
Weight(kg)	46.1±8.1	46.1±8.6	0.977
BMI	19.2±2.5	19.3±2.7	0.040

Table 7.3 depicts the distribution of anthropometric indicators in tea garden workers according to blood pressure status. Blood pressure, Age, BMI revealed a significant difference and have high values in hypertensive than normotensive.

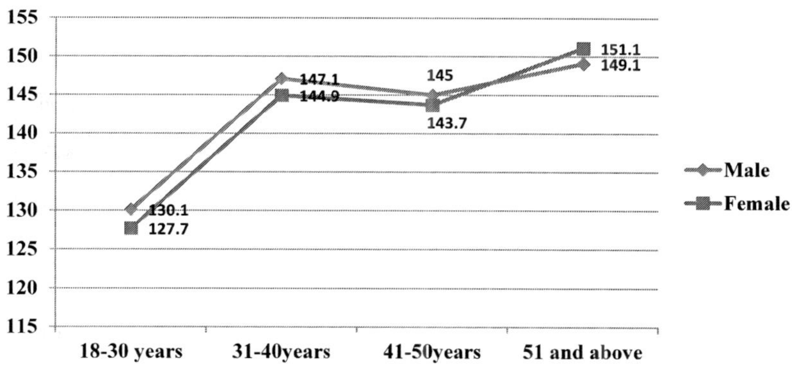

Figure: 7.1 Systolic blood pressure according to age category and sex

Figure 7.1 and 7.2 reveals that it increases in SBP and DBP in both genders. However, it was found that the rise of SBP and DBP at an age group of 51 and above among females than that of males (<0.001).

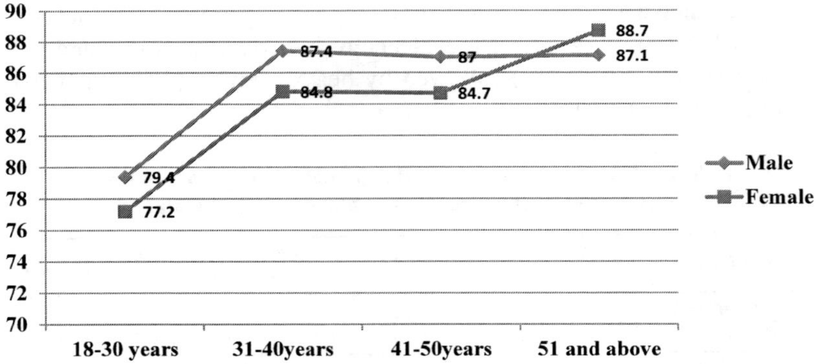

Figure.7.2 Diastolic blood pressure according to age category and sex

Table 7.4. Independent predictors of Hypertension determined by multivariate logistic regression analysis

Variables	*Subgroup*	*OR(95% CI)*
Age group (years)	18-30	Ref.
	31-40	3.8*(2.7-5.4)
	41-50	3.5*(2.3-5.5)
	51 and above	5.5*(3.1-9.5)
Sex	Female	Ref.
	Male	1.3(0.9-1.8)
Physical activity	Heavy	Ref.
	Light	1.4(0.9-2.2)
	Moderate	1.2(0.6-1.9)
Extra Salt	No	Ref.
	Yes	1.5*(1.1-2.0)
Smoking	No	Ref.
	Yes	1.2(0.7-1.9)
Alcohol consumption	No	Ref.
	Yes	0.9(0.6-1.3)
Chewing tobacco	No	Ref.
	Yes	1.8*(1.2-2.5)
BMI	Underweight (<18.5)	Ref.
	Normal (18.5 to 24.9)	0.8(0.3-1.9)
	Overweight (25-29.9)	0.2(0.1-1.5)
	Obese (> = 30)	0.8(0.6-1.1)

OR: adjusted odds Ratio, * significant (p <0.05)

Table 7.4 shows the odds ratios of different variables determined by multivariable logistic regression analysis. It is found that age, male gender [OR1.3, 95% CI (0.9-1.8)], consumption of extra salt [OR1.5, 95% CI (1.1-2.0)] and chewing tobacco [OR1.8, 95% CI (1.2-2.5)] is found to be an independent predictor of hypertension.

Discussion

The present study highlights hypertension and its behavioral risk factors among tea garden workers of Assam. The prevalence of hypertension was 51.9 percent and was higher among males (62.0%) as compared to females (38.0%) with significant sex differences. Similar studies like Mahanta *et al.,* (2008) found that the overall prevalence of hypertension was 33.3 percent with no significant sex difference. The overall prevalence of hypertension in their study was 63.63 percent 64.2 percent in men, and 62.89 percent in women which was highest and reported by Hazarika *et al.,* (2003).In the current study, the tea garden community having a habit of tobacco chewing, alcoholic users, taking extra salt were having hypertension. A similar study carried out by Mahanta *et al.,* (2013) found that tea-garden workers consumed more salt, tobacco consumption 85.2 percent ($p < 0.0001$).

This study shows that the test of significance records significant values in the anthropometric parameters like blood pressure, age (<0.001), body mass index (<0.005). A similar significant association was reported by Bendhari *et al.,* (2016), Midha *et al.,* (2009) ($p<0.0001$) reported that the distribution of study subjects according to Body Mass Index shows a strong and statistically significant association between obesity and prevalence of hypertension. However, the analysis of multivariate logistic regression shows that age category, extra salt, and chewing tobacco to be significantly associated with hypertension. A study carried out by Manimunda *et al.,* (2011) reported that there is a significant association between hypertension and various age categories and overweight/obesity ($p<0.05$). However, Bendhari *et al.,* (2016) found that age smoking, family history of hypertension found to be significant.

The present study has certain limitations. First, the assessment of smoking, alcohol, tobacco, age group was based on recall which may result in under-reporting of substance use. Secondly, the majority of tea garden workers were non-literate. Last, this study took a sampling from the sizeable population but results are considered to be sufficiently general enough those behaviour risk factors, access to health-care and dietary practices are similar.

Conclusion

From the present study, it is seen that the burden of hypertension was quite high (51.9%) among the tea garden community of Assam. Prevalence of substance use like tobacco, alcohol, and smoking are common among tea garden workers. Early implementation of health promotional measures including abstinence from tobacco consumption and moderation of salt intake will help to reduce the future risk of hypertension.

Acknowledgment

This study was an outcome of a completed project funded by the Indian Council of Medical Research (ICMR), New Delhi and the authors would like to acknowledge the Director of RMRC, Dibrugarh, Assam.

References

Bendhari, ML, Korade, RS & Haralkar, SJ 2016, 'Study of prevalence and risk factors of hypertension in adults in an urban slum area of Western Maharashtra, India'. *International Journal of Community Medicine and Public Health,* [online] vol.3, no.10, pp.2812-2816. Available from: DOI: http://dx.doi.org/10.18203/2394-6040.ijcmph20163366.

Hazarika, NC, Biswas, D & Mahanta, J 2003, 'Hypertension in the elderly population of Assam'. *JAPI.* [online] vol.51, pp.567-573. Available from: http://www.japi.org/june2003/O-567.pdf.

Hazarika, NC, Biswas, D, Narain, K, Kalita, HC & Mahanta, J 2002, 'Hypertension and its risk factors in tea garden workers of Assam'. *National Medical Journal of India*, [online] vol.15, no.2, pp.63-68. Available from: https://www.ncbi.nlm.nih.gov/pubmed/12044117.

Kann, L, Kinchen, SA, Williams, BI, Ross, JG, Lowry, R, Grunbaum, JA & Kolbe, LJ 2000, 'Youth risk behavior surveillance – the United States, 1999. State and local YRBSS Co-ordinators'. *Journal of School Health*, [online] vol. 70, no.7, pp.271-285. Available from: https://www.ncbi.nlm.nih.gov/pubmed/10981282

Kearney, PM, Whelton, M, Reynolds, K, Muntner, P, Whelton, PK & He, J 2005, Global burden of hypertension: analysis of worldwide data. *Lancet*, [online] vol. 365, no.9455, pp.217-223. Available from: https://www.ncbi.nlm.nih.gov/pubmed/15652604.

Mahanta, TG, Ahmed, FU, Mahanta, BN & Barua, A. 2008, 'Prevalence of hypertension and its risk factors in a tea garden community of Dibrugarh District, Assam'. *Indian Journal of Public Health*, [online] vol.52, no.1, pp.45-47. Available from: https://www.ncbi.nlm.nih.gov/pubmed/18700723.

Mahanta, TG, Joshi, R, Mahanta, BN & Xavier, D 2013, 'Prevalence of modifiable cardiovascular risk factors amongst tea garden and general population in Dibrugarh, Assam, India'. *Journal of Epidemiology and Global Health,* [online] vol.3, no.3, pp.147–156. Available from: doi:10.1016/j.jegh.2013.04.001.

Manimunda, SK, Sugunan, AP, Benegal, V, Balakrishna, N, Rao, MV & Pesala, KS 2011, 'Association of Hypertension with risk factors & hypertension related behaviour among the aboriginal Nicobarese tribe living in Car Nicobar Island, India'. *Indian Journal of Medical Research.* [online] vol.133, no.3, pp.287-293. Available from: http://www.ijmr.org.in/temp/IndianJMedRes1333287-4892744_133527.pdf.

Midha, T, Idris, MZ, Saran, RK, Srivastava, AK & Singh, SK 2009, 'A study on the association between hypertensive status and anthropometric correlates in the adult population of Lucknow District, India'. *Indian Journal of Preventive and Social Medicine.* [online] vol.40, no.1, pp.50-54. Available from: 0.18203/2320-6012.ijrms20171854.

World Health Report 2002, *Reducing risks, promoting a healthy life.* Geneva, World Health Organization, 2002.

World Health Organization 2013, *A global brief on Hypertension Silent killer, a global public health crisis.*

Chapter - 8

Anthropometric Somatotypes among the Pre-adolescent Ahom Children of Upper Assam

DALI DUTTA

Introduction

Sheldon and his co-workers coined the term 'Somatotype' (Sheldon *et al.*, 1940). The modified and developed version of Sheldon's technique is known as anthropometric somatotyping but it gained impetus in the last two decades following the introduction of a simplified method for somatotyping by Heath and Carter (1967), based on ten anthropometric measurements they forwarded a method of assessing anthropometric somatotype. Between the genders, worldwide variation exists in body size, body proportion, and body physique (Hall, 1982; Antoszewska *et al.*, 1992; Ghosh, 2004).

Somatotyping can be best thought of as a numerical shorthand method of describing the present morphological conformation in terms of body shape and composition independent of body size (Heath and Carter, 1967; Carter, 1975).

Considering the importance of understanding variation in body physique, the present study aims to study the pattern of somatotype and gender differences in anthropometric somatotype among the Ahom children inhabiting the districts of Upper Assam.

The Ahoms are one of the major Mongoloid populations of Assam. They belong to the Tai or Shan family and came to Assam from Myanmar through the Patkai range in 1228 A.D. Linguistically the Ahoms belong to the Siamese Chinese branch of the Sino-Tibetan language family. At present, they are mainly concentrated in Dibrugarh, Tinsukia, Lakhimpur, Dhemaji, Sibsagar, and Jorhat districts of upper Assam.

Material and methods

The study is cross-sectional. 586 pre-adolescent Ahom children aged 6 to 10 years of age were selected (282 boys and 304 girls). The present study was conducted between February 2010 to July 2011. All subjects were selected randomly. Consent was obtained from the parents and Head Master before the work commenced. The present Ahom boys data is compared with Tribal schoolboys of the West Tripura district of Tripura (Sarkar and Sil, 2014) and Ao Naga boys of the Mokokchung district of Nagaland (Longkumar, 2014). Due to a lack of data the girls' data can not be compared.

For the evaluation of the somatotype, the measurements were taken following internationally accepted standards. In addition to age, these 10 anthropometric measurements were: weight, height, upper arm and calf girths, triceps, subscapular, supra-iliac, and calf skinfolds, and bicondylar breadth of the humerus and femur. All the anthropometric measurements were taken on the right side of the subject, using Anthropometric Rod and Sliding Caliper. Circumferences were measured with the help of Steel Tape. All the skinfold measurements were taken on the left side of the subject, Skinfold measurements in the present study were taken with the Holtain skinfold caliper in mm, with the dial marked in the division of 0.2 mm.

A One-way Analysis of Variance (ANOVA) was used to examine sex differences in somatotype components. Statistical analysis were carried out using SPSS software programs.

Heath – Carter method was followed for somatotype rating. This method provides a three-number rating representing endomorphic, mesomorphic, and ectomorphic, respectively. The endomorphic component represented the adiposity of a body. Mesomorph components represented muscularity. The third somatotype component, the ectomorphy expressed the relative linearity of the body. Values between 0.5 and 2.5 are considered low, between 3 and 5 moderate, between 5.5 and 7 high, and above 7.5 very high.

The following equations were used for calculating somatotype.

Endomorphy = % $0.7182 + 0.1451 \times \Sigma SF\ 0.00068 \times \Sigma SF2 + 0.0000014 \times \Sigma SF3$, where ΣSF = (sum of triceps, subscapular, and supra-iliac skinfolds) multiplied by (170.18/height in cm). This was called height-corrected endomorphy and was the preferred method for calculating endomorphy)

Mesomorphy = 0.858 × humerus breadth + 0.601 × femur breadth + 0.188 × corrected arm girth + 0.161 × corrected calf girth % height × 0.131 + 4.5.

Three different equations were used to calculate ectomorphy according to the height-weight ratio (HWR):

(i) If HWR was greater than or equal to 40.75 then, Ectomorphy = 0.732 × HWR % 28.58

(ii) If HWR was less than 40.75 and greater than 38.25 then, Ectomorphy = 0.463 × HWR % 17.63

(iii) If HWR was equal to or less than 38.25 then, Ectomorphy = 0.1

Results and discussion

Table 8.1 shows the mean somatotype and height weight ratio (WHR values for Ahom children according to age. It is relevant from the Table that at the age of 6 years the mean somatotype of Ahom boys is 1.58-4.65-2.74 and in the age 10 it is 1.40-4.02-4.11. this shows that the endomorphy and mesomorphy of boys are decreases by 0.18 and 0.63 units respectively. But the ectomorphy is increased by 1.37 units. The girl's somatotype in age 6 years 1.49-4.41-2.73 and 1.65-3.65-4.08 at the age of 10 years. In the girls, the endomorphy and ectomorphy are increased by 0.16 and 1.35 units respectively. On the contrary, a mesomorphy increase of 1.35 units. For both the sexes the Waist Hip Ratio shows increasing trends concerning age. This evidence supports that as age progresses there is a change in somatotype ratings.

Table 8.1: Mean somatotype and height weight ratio values for Ahom children according to age

Age (years)	*Sex*	*Number*	*Somatotypes (Mean ±SD)*	*Waist Hip Ratio (Mean± SD)*
6	Boys	67	1.58-4.65-2.74 (0.40-0.91-1.12)	42.71 (1.72)
	Girls	57	1.49-4.41-2.73 (0.32-0.86-1.06)	42.75 (1.50)
7	Boys	53	1.18-4.02-3.65 (0.39-0.75-1.10)	44.03 (1.50)
	Girls	60	1.39-3.99-3.38 (0.43-0.67-0.92)	43.66 (1.25)

(Cond...)

8	Boys	54	1.31-3.84-3.99 (0.33-0.71-1.09)	44.50 (1.49)
	Girls	54	1.45-3.70-4.03 (0.45-0.78-0.87)	44.55 (1.18)
9	Boys	55	1.54-3.90-4.03 (0.44-0.90-1.13)	44.56 (1.54)
	Girls	64	1.49-3.65-4.13 (0.47-0.73-0.89)	44.68 (1.22)
10	Boys	53	1.40-4.02-4.11 (0.39-0.72-1.04)	44.66 (1.42)
	Girls	69	1.65-3.65-4.08 (0.58-0.85-1.07)	44.61 (1.46)

One-way ANOVA revealed that sex differences are not statistically significant (P<0.05) for all the three components of somatotype, i.e. Endomorphy, Mesomorphy, and Ectomorphy, among Ahom (Table 8.2). Variation as evident from the co-efficient of variation is lowest for endomorphy among both males and females.

Table 8.2: Sex differences in somatotype components of Ahom (One-way analysis of variance)

ANOVA

		Sum of Squares	*df*	*Mean Square*	*F*	*Sig.*
Endomorphy	Between Groups	5.823	4	1.456	7.663	.000
	Within Groups	108.469	571	.190		
	Total	114.292	575			
Mesomorphy	Between Groups	52.084	4	13.021	20.314	.000
	Within Groups	366.002	571	.641		
	Total	418.086	575			
Ectomorphy	Between Groups	165.319	4	41.330	38.831	.000
	Within Groups	607.739	571	1.064		
	Total	773.058	575			

Somatotypes of pre-adolescent Ahom boys and girls are classified into thirteen categories following Carter's classification (Carter, 1980) and presented in table 8.3. Ahom boys and girls are predominantly Ectomorphic Mesomorph 70.92 percent and 69.08 percent respectively, followed by, Mesomorphic Endomorph 14.18 percent and 14.80 percent and Ectomorphic-Endomorph 10.64 percent and 9.87 percent A very small section of girls belongs to Endomorphic-Ectomorph (0.99%) and Balanced Mesomorph (0.33%). Only 0.35 percent of boys belong to Mesomorphic Ectomorph.

Table 8.3: Frequency distribution of Ahom children in various somatotype categories

Somatotype categories	*Boys*		*Girls*	
	No	*%*	*No*	*%*
Mesomorphic Endomorph	40	14.18	45	14.80
Mesomorph- Ectomorph	1	0.35	6	1.97
Endomorphic Mesomorph	7	2.48	5	1.64
Balanced Mesomorph	1	0.35	1	0.33
Ectomorphic Mesomorph	200	70.92	210	69.08
Mesomorphic-Ectomorph	3	1.06	4	1.32
Balanced Ectomorph				
Endomorphic Ectomorph			3	0.99
Endomorph-Ectomorph				
Ectomorphic Endomorph	30	10.64	30	9.87
Balanced Endomorph				
Central				
Total	282	99.98	304	100.00

The present sample of Ahom boys has been compared with samples (Table 8.4) of boys of similar age range derived from Meitei boys of Imphal district, Manipur (Gaur *et al.,* 1999), Tribal schoolboys of West Tripura district of Tripura (Sarkar and Sil, 2014) and Ao Naga boys of Mokokchung district of Nagaland (Longkumar, 2014). The present Ahom boys are generally Ectomorphic Mesomorph, whereas the tribal school boys and Ao Naga boys are dominantly Mesomorphic.

Table 8.4: Comparison of somatotype of Ahom boys with different population groups of North-east India

Age groups	*Tribal schoolboys of Tripura*	*Ao Naga of Nagaland*	*Meitei boys of Manipur*	*Present study*
6	-	-	2.0 - 3.6 - 3.8	1.58-4.65-2.74
7	-	-	1.9 - 3.9 -3.7	1.18-4.02-3.65
8	1.60-4.90-3.36	1.76-4.14-3.06	1.9 - 4.0 - 3.6	1.31-3.84-3.99
9	1.78-4.03-3.73	1.87-4.08-3.19	1.8 - 3.8 - 3.9	1.54-3.90-4.03
10	1.75-4.28-3.37	1.85-4.18-3.60	1.7 - 3.3 - 4.4	1.40-4.02-4.11

From the above results, it can be concluded that the Ahom children are predominantly Ectomorphic Mesomorph. The higher rating of Ectomorphic and Mesomorphic may be due to their diet and active participation in physical activities. As the present study is examining the anthropometric somatotype of pre-adolescent Ahom children, more research would be helpful to give a conclusion.

References

Antoszewska A, & Wolanski, N. 1992. Sexual dimorphism in newborns and adults. *Stud Hum Ecol.* 10: 23-38.

Carter JEL 1975. *The Heath-Carter Somatotype Method.* San Diego: State University Syllabus Service.

Carter JEL. 1980. *The Heath and Carter Somatotype Method.* 3rd edition. San Diego State University Syllabus Service, San Diego.

Gaur, R, Singh, SG & Lakhanpal, M 1999, 'Somatotypes of urban Meiteis of Imphal, Manipur', *Anthropologist*, vol.1, no.4, pp. 235-240.

Ghosh, A. 2004, 'Age and sex variation in measures of body composition among the elderly Bengalee Hindus of Kolkata, India', *Collegium Antropologicum,* vol. 28, no.2, pp. 553-561.

Hall, RL 1982, *Sexual Dimorphism in Homo sapiens: A Question of Size.* Praeger Scientific. New York.

Heath, BH & Carter, JEL 1967, 'Growth and somatotype of Manus children, Territory of Papua and New Guinea: Application of a modified somatotype method to the study of growth pattern', *American Journal of Physical Anthropology*, vol.19, pp.173-184.

Longkumer, T, 2014, 'Physical activity and somatotypes among Ao Naga boys', *Anthropologists,* vol.17, no.2, pp.669-675.

Sarkar, SR & Sil, SK 2014, 'Somatotype of non-athlete tribal schoolboys of West Tripura District, Tripura', *Biolife,* vol. 2, no.4, pp.1365-1370.

Sheldon WH, Stevens SS, and Tucker WB. 1940. *The varieties of human physique.* Harper Bros. New York.

Chapter - 9

Food Security and Coping Strategies among the Mishing Residing in flood-affected regions of Lakhimpur District, Assam

Nitamoni Bharali and Nitish Mondal

Introduction

Food and nutrition security is a basic human need and essential for living a healthy and productive life and considered to be fundamental to socio-economic development. Food security in a household could be said as 'Access by all people at all times to enough or sufficient foods for an active and healthy life'. The most widely accepted definition of food security is given by the Food and Agriculture Organisation of the United Nations (FAO) that 'A situation that exists when all people, at all times, have physical, social and economic access to sufficient, safe and nutritious food that meets their dietary needs and food preferences for an active and healthy life' (FAO *et al.,* 2014). Food security and agricultural production are considered to be the leading concerns associated with climate change in population (Schmidhuber and Francesco, 2007; Parry *et al.,* 2009; World Bank, 2010; Dasgupta *et al.,* 2012; Chakrabarty, 2016; FAO, 2016; Bhargava, 2019). It is evident that climate change effects and stresses food security and hunger in complex ways and it impacts crops, livestock, forestry, fisheries, and aquaculture, and can cause grave social and economic consequences in the form of reduced incomes, eroded livelihoods, trade disruption and adverse health impacts in individuals/population (Gregory *et al.,* 2005; FAO, 2014; Chakrabarty, 2016; Bhargava, 2019). Climate change may affect food systems in several ways ranging from direct effects on crop productions include the changes in rainfall leading to drought or flooding, or warmer or cooler temperatures leading to changes in the length of the growing season to changes in markets, food production, and prices and supply chain infrastructure (Gregory *et al.,* 2005; FAO, 2014; Bhargava, 2019). The prevalence of food insecurity and hunger in a household is referring as limited or uncertain availability of nutritionally adequate and safe foods or limited or uncertain ability to acquire acceptable foods in socially acceptable ways of sufficient foods. Moreover, the prevalence of undernutrition and/or malnutrition is caused by an assortment of interrelated factors including household food insecurity, poor household care of women and children, and poor availability or access to healthcare and sanitation in the population (Chakrabarty, 2005; Black *et al.,* 2008; McDonald *et al.,* 2015; Longvah *et al.,* 2017; Mastiholi *et al.,* 2018; Bharali *et al.,* 2019).

Food insecurity is a condition where a lacks of access to sufficient quantity and quality of food for a healthy and active life, which can compromise the overall health and nutrition status, wellbeing, and productivity of an individual or population. A country with many food insecure populations can even lead to a lower Gross Domestic Product for a country, making food insecurity an economic challenge and as well as a human rights problem (FAO 2012; White and Masset 2007; Jones *et al.,* 2013). Several research investigations have reported the higher magnitude of food insecurity and hunger among a nutritionally vulnerable segment of the population in India (Agarwal *et al.,* 2009; Pasricha *et al.,* 2009; Gopicharan *et al.,* 2010; Dasgupta *et al.,* 2012; Chinnakali *et al.,* 2014; Mastiholi *et al.,* 2018). Flood is a regular and very vicious problem of Assam. A previous investigation has stated that as the food-insecure status in a household becomes more severe, more coping strategies are adopted by its household members (Gupta *et al.,* 2015). The prevalence of hunger is a potential, although not necessary, consequence of food insecurity in which all people at all times have both physical and economic unable to access sufficient food to meet their dietary needs for a productive and healthy life. Hence, the existence of food insecurity and hunger is a challenge to human society, affecting the physiological, environmental, and economic development of the community or population. Therefore, to address this challenge, there is a need for comprehensive information on the nature and prevalence of food security, and also a proper identification of the people affected concern. This will allow the government agencies and policymakers to formulate more effective/target specific policies, programmes, and food aid design and monitoring in population, but the multi-dimensional nature of food security has been a challenge as most indicators only measure one dimension and attempt to create a composite indicator have thus, far not been successful in community/population. Despite the importance of the nutritional status and food system of the indigenous population, the area has received very little attention from the populations of North-east India.

Aims and objectives of the present investigation

This present investigation, thus, aims to assess food insecurity and coping strategies during the flood season through a qualitative approach among the indigenous Mishing population of Assam, North-east India. Identifying the coping strategies of Mishing peoples during the flood season when they have a limited amount of food will provide a better understanding of the severity of their condition and may assist in developing appropriate and effective programmes addressed for Mishing people households.

Material and methods

The present investigation was carried out among 50 households of the Mishing individuals of Assam, North-east India. The Mishing ethnic populations are the second largest tribal population in Assam. They are generally considered as an endogamous tribal population belonging to the Indo-Mongoloid population and the Tibeto-Burman linguistic family (Taid, 1987; Kakoti, 2017). The Lakhimpur is a district is situated in the Northern part of Assam and its total area covered 2,277 km^2 and having a population of 1,042,137 (Male- 5,12,463, Female-5,29,674) and a literacy rate of 77.20 percent (Male- 83.52%, Female- 70.62%) as per Census (2011). The investigation was carried out among homogenous Mishing villages of Khaboli Ghat and Pahumara area of Lakhimpur district of Assam (26°48'N and 27°53'N and 93°42'E and 94°20'E). A total of 166 adults (Male-80; Female-86) and 90 children belonging to the indigenous Mishing population were covered. The selection of the subject was done by using their place of residence, cultural background, surnames, and official records. The ethnicity of the research participants was identified by the identity cards, voter identity and ration cards, and relevant official records issued by the local government authority. The stratified random sampling method was used to include the research participants in the present investigation. The households were selected using a stratified random sampling method. The Mishing women and heads of the households of the respective households were interviewed to collect the relevant data. The data of the present investigation were mainly cover by household visits using an in-depth-interview method. The mixed-method design approaches which include both quantitative and qualitative methods were used to record the data in the present investigation. The food insecurity and hunger data were analyzed by a standard scale proposed by

Table 9.1: The levels of food insecurity in the Mishing population

Sl No.	*Food Insecurity Status (For the past four weeks)*			*Often <10 time*	*Sometimes (3-10 times)*	*Rarely (1-2 time)*	*During Which time*		
							Breakfast	*Lunch*	*Dinner*
1.	Moderate Food Insecurity Did you worry that your household would not have enough food?	Y	N						
2.	Were you or any household member not able to eat the kinds of foods you preferred because of a lack of resources?	Y	N						
3.	Did you or any household member have to eat a limited variety of foods due to a lack of resources?	Y	N						
4.	Severe Food Insecurity Did you or any household member have to eat some foods that you did not want to eat because of a lack of resources to obtain other types of food?	Y	N						
5.	Did you or any household member have to eat a smaller meal than you felt you needed because there was not enough food?	Y	N						
6.	Did you or any household member have to eat fewer meals in a day because there was not enough food?	Y	N						
7.	Food Insecurity with Hunger Was there ever no food to eat of any kind in your household because of a lack of resources to get food?	Y	N						
8.	Did you or any household member go to sleep at night hungry because there was not enough food?	Y	N						
9.	Did you or any household member go a whole day and night without eating anything because there was not enough food?	Y	N						
10.	Do you beg and borrow food from other people	Y	N						

Source: Blumberg, 1999.

Blumberg (1999). The level of food insecurity was categorized at various levels i.e., Moderate Food Insecurity, Severe Food Security, and Food Insecurity with Hunger showing in tables 9.1 and 9.2. The samples of questions were asked for conducting the in-depth interview on the Mishing women of the flood-affected area during and/or after the occurrence of the flood seasons.

Table 9.2: A sample questionnaire for in-depth interview to collect the food insecurity among Mishing population

In-Depth Interview questions
• Your typical/usual eating pattern (seasonal food, seasonal period)? • Kinds of food available in your home in the last 3 months? • Whether happy or not with the food available at home? • Did it make any difference in your life? How? • Are there any ways that you could try to help your family? • The ways that your family gets food? • Who is the most responsible person for household food allocation and food management? • Does your family have difficulty in providing food? • How do you and your family manage that lack of food? • In your opinion, is your food enough for you? Why? • Have you felt a lack of food at home in the last 3 months (the Flood season period)? YES/NO • How did you handle it? • Do you ever get food from another source that your family cannot access or the other way around? YES/NO • Can you tell me about a time when you had to do something if they didn't have food? • Traditional System of preserving food to cope up with flood? • What are Government Scheme, Local, College, NGOs relief Material provided to them? Whether individuals are satisfactory or not? If not why? And what are their demands?

Socio-economic, demographic and lifestyle variables

The socio-economic demographic data including age, sex, monthly family income, sanitation, dependent children and parental education, house type, water facility, electricity, the number of living rooms, dependent children, occupation, and household income will be recorded for each household, monthly investment for food, Invest during flood, land owned, land affected by natural calamities, property lost during natural calamities or flood. A pre-structured and pre-tested open-ended schedule/questionnaire was completed by the household visits and interviewing the subjects or respondents. The socio-economic status was evaluated using a modified version of the scale of Kuppuswamy (Mishra and Singh, 2003; Kumar *et al.,* 2007). The scale determines the socio-economic based on a score calculated from education, occupation, and monthly income. The occurrence of non-communicable and communicable diseases is also being recorded. The households included in the present investigation are belonging to lower to middle socio-economic groups using the modified scale of Kuppuswamy. Furthermore, the participation of the research participants in the present research investigation was purely voluntary, and informed consent was obtained before the inclusion and data collection. The objectives, research goals of the present investigation, and nature of participation were explained to the research participant before the collection of the relevant data. The present research investigation was carried out according to the ethical guidelines for human research as laid down in the Helsinki Declaration (Touitou *et al.,* 2004; Portaluppi *et al.,* 2010).

Statistical analysis

The data of the present was done using both Statistical Package for Social Sciences (SPSS, 16.0) and MS-Excel for Windows. First, standard univariate descriptive statistics and frequency distributions were calculated. The bivariate analysis of food insecurity and hunger combined with and without sex among households examined by cross-tabulation. P-values of <0.05 were considered statistically significant.

Results

The Mishing of Lakhimpur district facing several problems includes lack of education, lack of healthcare facilities, and ignorance by the left of the society, various social-economical environmental and geographical dispossessions. The Mishing peoples are the dweller of the riverside as they were very much prone to flood during the monsoon season. It is a regular phenomenon that Mishing people dealing with for many years. During the extreme flood, food insecurity and hunger were found to be is much high up among the Mishing individuals of these flood-affected areas. The results reveal that more than 90.0 percent of the studied population were affected by the flood caused by the river Ranganadi and Subansiri of Assam. Therefore, several coping strategies were found to be adopted by Mishing households of flood-affected areas. An interview discloses Mishing people's general experience during and/or after the flood period. All the households stated that they frequently experienced hunger, especially during the flood season. The flood season occurs mainly between July and September every year. The flood season affects their food availability because they rely on their food. Flood-damaged in their fields and their home-preserved food or it's also hampering further cultivation as it's damaged the seeds and cultivated land by leaving sands or wastage on the ground.

Table 9.3. Food insecurity status among the Mishing household members of Lakhimpur district of Assam

Individuals	*No food insecurity*	*Moderate Food Insecurity*	*Severe Food Insecurity*	*Food Insecurity with Hunger*	*Total*
Male	0	1 (1.25)	18 (22.5)	61 (76.25)	80 (31.25)
Female	0	1 (1.16)	15 (17.44)	70 (81.39)	86 (33.59)
Children	0	4 (4.44)	50 (55.55)	36 (40.00)	90 (35.15)
Total	0	5 (1.93)	83 (32.42)	167 (65.23)	256 (100.00)

Figures within parentheses indicate percentages

The result was found that the women of the Mishing household were more severely affected by Food Insecurity including hunger but on the other hand male members of the household were found to be more affected by moderate or severe food insecurity (Food Insecurity with Hunger- Male: 76.25 % vs. Female: 81.39%; Severe Food Insecurity- Female: 17.44% vs. Male: 22.50%; Moderate Food insecurity- Male: 1.25% vs. Female: 1.16%). As well as the children were also found to be severely affected by household food insecurity (Food Insecurity with Hunger- 65.23%, Severe Food Insecurity-55.55%, Moderate Food insecurity-1.93%). Mishing household was absent found to have food security after flood seasons of the Lakhimpur district of Assam. Overall, it was noticed that all the individuals include males, females, and children of the Mishing households were suffering from food insecurity with hunger at a very high level in the flood seasons.

Coping strategies among Mishing population

Several coping strategies were applied and/or adopted by the Mishing people to cope with the situation caused due to flood includes self-initiated coping strategies implemented at the individual level. The listed coping strategies are as follows:

- Eating less-desirable food.
- Reducing portion size of the food.
- Eating rice only, without vegetables or side dishes.
- Skipping one or two meals at times.
- Borrowing food from neighbours or others.
- Selling own field produce and livestock.
- Strategies included relying on less expensive foods like seasonal and/or locally available vegetables.
- Limiting the quality of meals and reducing numbers of healthy or nutritious foods.
- To increase the short-term availability of foods and preserved food for flood season and breeding and production for the next seasons.
- In the case of extreme insecurity, they only depend on relief materials supplied or provided by the Government and other sources.

Amongst all the above coping strategies they adopted it was to be noted that the most preferable one is relay on the relief materials given by various Governments, non-Government agencies, and other sources. The second most applied strategy includes preservation methods of foods and grains reported by the Mishing populations. They preserved food for flood season by various traditional preservation methods and it could be also used for further breeding. After that, the most adopted coping strategy was like limiting quality meals and reducing numbers of healthy food, eating less-desirable food, reducing the portion size of the food, eating rice only, without vegetables or side dish and relying on less expensive foods like seasonal or locally available vegetables. Moreover, it is being reported that due to extreme flood effects, they use to skip meals one or two times a day.

The frequencies of different responses of coping strategies are depicted in figure 9.1. The additional problem that arises after the flood is like the people who have been sheltered in the broken embankments and highland areas of the flood-affected regions. Due to this, a very congested way of settlement pattern has been found inside the embankments in the studied areas. Coping strategies include the preparation of the local country boat for future flood disasters. The Mishing people store food and save money for difficult times and try to keep themselves informed about flood situations and warnings as preparatory measures and sometimes using their folk beliefs and observational experiences. Transportation during floods is usually on rafts made from the wood of the banana tree and they help each other in rescue and evacuation operations. Reductions of land and forest resources due to floods have been causing problems in the areas. Reduction of land use has been recorded due to sand casting after the result of embankment being breached by floodwaters due to this the field and the vegetable become

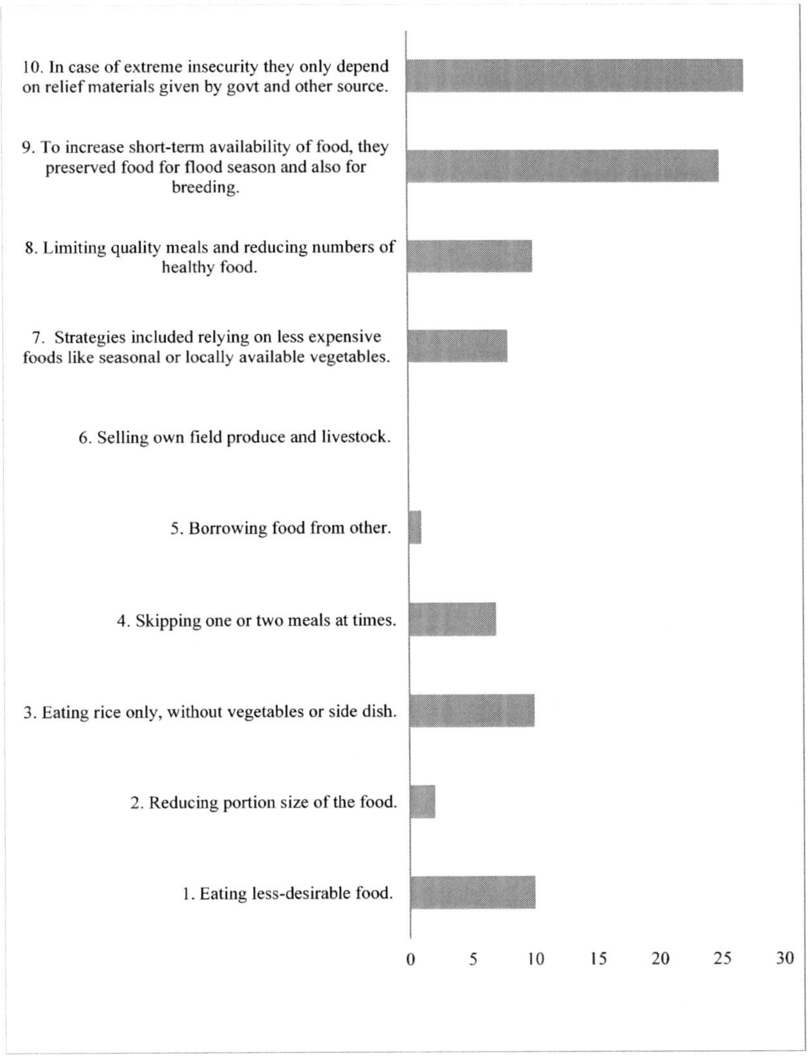

Figure 9.1: Frequencies of different coping strategies adopted by the Mishing households during the extreme flood seasons

decreasing. The Mishing community of the district has been very badly affected by sand casting, soil erosion/riverbank erosion, destruction of their houses and natural surroundings, and paddy field vegetable production and livestock. These factors result in the unavailability of food production, damage of preserved food materials, scarcity of food during an extreme flood situation. Therefore, food insecurity and hunger are becoming an obstacle in the lifestyle of the Mishing populations dwelling on the riverside. Furthermore, it is observed that the reduction of meal size or skip meals and often follow up were combined into a single item and this allows coding of the food items for recurring cutting/skipping, which is more consistent with food insecurity by hunger. The research participants included were mainly belonging to low to middle socio-economic status using the modified scale of Kuppuswamy's and overall the consumption of food was observed to be insufficient or inadequate due to non-availability and affordability of the rural Mishing population due to occurrence of flood in the regions. Therefore, identifying such coping strategies by the Mishing household during the flood season when they have a limited amount of food will provide a better understanding of the severity of their conditions and may assist in developing appropriate and effective programmes addressed for Mishing people households.

Discussion

The prevalence of food insecurity is considered to be a complex and multi-dimensional phenomenon that is not appropriately evaluated by the currently recommended and used as conventional proxy indicators to determine the social problem in several developing countries including India. Moreover, several indigenous/tribal ethnic communities are belonging to the vulnerable segments and also residing in rural regions that mainly depend on smallholder farming faced food insecurity induced due to climate-related disasters (drought or flood). In response, several ethnic communities/ populations are taking the initiative to cope up and adapt to climate-related disasters in households/ individuals level in the population (Gupta *et al.*, 2015; Ali and Vallianatos, 2017; Farzana *et al.*, 2017). India's North-eastern state of Assam has been hard-hit by monsoon rains and flooding in the past and present. The flood is submerged a huge segment of human habitat by the rising waters of the river Brahmaputra and its tributaries in Assam and other north-eastern states in India. Therefore, the individuals belonging to these regions were much affected by the flood, or the hazardous effects caused by the flood are attributed to food insecurity, loss of food and property, and human migration in the population.

Food insecurity and hunger are considered to be the regular humanitarian conditions among the tribal/indigenous population habitat in areas of flooded rivers planes. Therefore, the seasonal monsoon amplifies the economic drivers of food insecurity and hunger among Mishing populations (Table 9.3). Agarwal *et al.,* (2009) reported the prevalence of food insecurity of 51.0 percent in urban Delhi. The prevalence of food security was observed to high in Coimbatore, Karnataka (58.0%, Pasricha *et al.,* 2009), Mizoram (58.0%, Medecins Sans Frontieres Foundation, 2008), and Khasis of Meghalaya (80.0%, Chyne *et al.,* 2017). Gopicharan *et al.,* (2010) have reported the high prevalence of food insecurity (74.6%) among the urban Tamil Nadu population in India. Studies have confirmed that populations belonging to particularly vulnerable segments residing in rural areas had home to small and marginal farmers who rely on rain-fed mono-cropping, which provides barely a few months of food security in a normal year in India (Dasgupta *et al.,* 2012; FAO, 2016). Ramachandran (2014) reported that food stocks begin to run out three or four months after harvest, farm jobs are unavailable and by the next monsoon/sowing season, food shortages peak to hunger. Seasonal floods will also hurt the livelihoods of live-stocks and agriculture-dependent individuals. Landless agricultural labourers wholly dependent on agricultural wages are at the highest risk of losing their access to food (Schmidhuber and Tubiello, 2007; Chakraborty, 2016). The whole family has to suffer from food insecurity like adults and/or children in households with moderate food insecurity reported multiple indicators of problems of food access such as inadequacy in household food supplies or adjustments to the quality of food consumed. In addition to these indicators, households with severe food insecurity also reported disrupted eating patterns and reduced food intake among adults and/or children (World Bank, 2010). Researchers have also confirmed that the extent to which food insecurity and hunger are occasional, recurring, or frequent in India (Chakravarty, 2005; Ramakrishna and Gadam, 2014). On the other hand, the Mishing people had to suffer from various kinds of health-related issues causing by floodwaters like skin disease, diarrhea, and other severe health complications.

Moors *et al.,* (2013) reveal the harmful effects of waterborne diseases like diarrhea in northern India due to flood water and long-time effects on the population (particularly children) are like undernutrition (Islam *et al.,* 2014; Bharali *et al.,* 2019). Moreover, the prevalence of malnutrition/undernutrition is considered to be the most serious consequence of food insecurity and has a magnitude of adverse health consequences and economic implications in the population (Chinnakali *et al.,* 2014). The poor attainment of physical growth pattern, a high magnitude of undernutrition and anemia was observed among the indigenous/ethnic tribal populations belonging to the vulnerable segments of the populations of North-east, India (Basu *et al.,* 2012; Sikdar, 2012; Islam *et al.,* 2014; Chyne *et al.,* 2017; Longvah *et al.,* 2017; Bharali and Mondal, 2018; Mastiholi *et al.,* 2018; Bharali *et al.,* 2019).

The necessary improvement of the system includes food production, food distribution, and economic access may all contribute to food systems adapted to cope up with climate change, but in adopting such changes it will be important to ensure that they contribute to sustainability (Gregory *et al.,* 2005). Moreover, despite advances in food production and distribution technologies, global food insecurity continues in developing countries including India (Agarwal *et al.,* 2009; Pasricha *et al.,* 2009; Gopicharan *et al.,* 2010; Ali and Vallianatos 2017). In connection to food insecurity, the adoption, and execution of new techniques and/or alteration of regular behaviors that translate to coping strategies in individuals/populations (Ali and Vallianatos, 2017; Farzana *et al.,* 2017). The low ability of the food to cope with shocks and to mitigate long and/or short-term stresses means that coping strategies might be available in other regions are unavailable or inappropriate in the population (Gregory *et al.,* 2005). The coping strategies are an indicator of food security and they differed from one person to another due to their environmental conditions and/or encountered nature of disaster or natural calamity. Several researchers have reported that managing household coping strategies revealed the consumption of the limited varieties of food available at home could indicate the severity of the coping strategies (Gregory *et al.,* 2005; Gupta *et al.,* 2013). Mishing individuals/population adopted several self-initiated coping strategies before the reaching of any other external help examples include a severe reduction in food consumption, eating non-food items or low food value item, borrowing money, taking debt with high interest, reducing expenditures on basic services such as health and education, a reduction in the number of consumers in the family sending to live and or work elsewhere (Figure 9.1). It was observed that a household adopts both positive and negative coping strategies simultaneously to overcome food insecurity among Mishing individuals. The execution of coping strategies significantly differed among female-headed households than male-headed and the infrastructure of the households. Moreover, accurately identifying and assessing the experience of food security in individuals and/or populations were considered to be essential for the development and implementation of effective programmes to solve problems related to food insecurity, poor dietary intake, and low nutritional status as a step towards achieving overall health status by reducing morbidity and health consequences in the population (Gregory *et al.,* 2005; Belachew *et al.,* 2011).

Strengths and limitations of the present investigation

The nutritional status and food security of the indigenous tribal population are of prime interest owing to their ethnic, cultural, socio-economic, and geographic variations in populations. The data of the present investigation were obtained through face-to-face in-depth interviews and there were no missing values amongst the subjects/research participants. The present investigation is cross-sectional and restricting on small sample size for the purpose and the scale assessment statistics indicated good reliability implying that the questions were well framed and that respondents have given ample attention to the questions, understood them, and sought thoughtful responses during the present investigation.

Conclusion

Food security has to be addressed using a multi-sectoral approach to mitigate the disastrous impacts of floods. The inter-linkages between various socio-economic factors have to be addressed to assess the impact of floods on food insecurity among Mishing individuals. The findings revealed that incessant rains and sudden floods had disastrous impacts on Mishing households and the floods have affected the infrastructure badly by damaging roads and bridges and impacted negatively the delivery of services such as health, agriculture, and education in the study area. Agriculture and livestock got severely affected during floods. The food security situation got worsened and the floods have impacted the social systems and food habits of the households. Moreover, the individuals/households who have difficulties consuming an adequate amount of food had to implement some coping strategies to overcome food insecurity. Understanding this situation may be useful for planning better-targeted food insecurity alleviation programmes, especially involving vulnerable groups such as children, adolescents, and women, to reduce the number of food-insecure individuals as well as food-insecure households and, thus, to create a healthier future generation. Therefore, the government needs to take a long-term view of disaster relief strategies. Moreover, given the adverse impacts of natural disasters on child nutrition, long-term undernutrition prevention programs must be implemented in disaster-affected regions. Additional efforts must be directed towards reducing the risk in agriculture. Such schemes should be specially targeted towards small farmers.

Acknowledgement

The authors acknowledged the help and co-operation of the people of the study area, parents/authority of all the households, local-level community leaders, and village headmen during the fieldwork. The extended help and co-operation of the Department of Anthropology, Assam University (Diphu Campus) is also acknowledged.

References

Agarwal, S, Sethi, V, Gupta, P, Jha, M, Agnihotri, A & Nord, M 2009, 'Experiential household food insecurity in an urban underserved slum of North India', *Journal of Food Security*, vol. 1, pp.239-250.

Ali, HMA & Vallianatos, H 2017, 'Women's experiences of food insecurity and coping strategies in the Chittagong Hill Tracts, Bangladesh'. *Ecology of Food and Nutrition*, 56:462-478. doi: 10.1080/03670244.2017.1381604.

Basu, D, Gulenur Islam, G, Gogoi, R, Dey, S & Deori, J 2012, 'Child's growth and nutritional status in two communities-Mishing tribes and Kaibarta caste of Assam, India', *International Journal of Sociology and Anthropology*, vol.6, pp.59-69.

Belachew, T, Hadley, C, Lindstrom, D, Gebremariam, A, Michael, KW, Getachew, Y, Lachat, C & Kolsteren, P 2011, 'Gender differences in food insecurity and morbidity among adolescents in south-west Ethiopia', *Journal of Pediatrics*, 127:e398-405. doi: 10.1542/peds.2010-0944.

Bharali, N & Mondal, N 2018, 'Socio-economic correlates and prevalence of Undernutrition among Nyishi Tribal Pre-school Children of North-east India', *South Asian Anthropologist*, vol.18, pp.47-57.

Bharali, N, Singh, KN & Mondal, N 2019, 'Composite index of anthropometric failure (CIAF) among Sonowal Kachari tribal pre-school children of flood a effected region of Assam, India', *Anthropological Review*, vol.82, pp.163-176.

Bhargava, A. 2019, Climate change, demographic pressures, and global sustainability. *Economics and Human Biology*, vol.33, pp.149-154. doi: 10.1016/j.ehb.2019.02.007.

Black, RE, Allen, LH, Bhutta, ZA, Caulfield, LE, de Onis M, Ezzati M, Mathers C, Rivera J & Maternal and Child Undernutrition Study Group 2008, 'Maternal and child undernutrition: global and regional exposures and health consequences', *Lancet*, vol. 371, pp.243–260.

Blumberg, SJ, Bialostosky, K, Hamilton, WL & Briefel, RR 1999, 'The effectiveness of a short form of the household food security scale', *American Journal of Public Health*, vol.89, pp.1231-1234.

Chakrabarty, M 2016, Climate change and food security in India. ORF 157:1-12.

Chakrabarty, S 2005, *Food insecurities in India: Cause and dimensions*. Indian Institute of Management, Ahmedabad. (http://hdl.handle.net/11718/6274)

Chinnakali, P, Upadhyay, RP, Shokeen, D, Singh, K, Kaur, M, Singh, AK, Goswami, A, Yadav, K & Pandav, CS 2014, 'Prevalence of household-level food insecurity and its determinants in an urban resettlement colony in north India', *Journal of Health Population and Nutrition*, vol.32, pp.227-236.

Chyne, DAL, Meshram, II, Rajendran, A, Kodali, V, Getti, N, Roy, P, Kuhnlein, HV & Longvah, T 2017, 'Nutritional status, food insecurity, and biodiversity among the Khasi in Meghalaya, North-east India', *Maternal and Child Nutrition,* Suppl 3. doi: 10.1111/mcn.12557.

Dasgupta, S. Roy, S & Sarraf, M 2012, 'Urban flooding in a changing climate: Case Study of Kolkata, India', *Asian–African Journal of Economics Econometrics,* vol. 12, pp.135.

FAO, IFAD & WFP 2014, The state of food insecurity in the World 2014. *Strengthening the enabling environment for food security and nutrition*. Rome. Available at: http://www.fao.org/ publications/sofi/2014/en/.

FAO 2012, *The State of Food Insecurity in the World 2012* Key messages, Rome: FAO

FAO 2014, *The State of Food Insecurity in the World the multiple dimensions of food security*. Key messages, Rome.

FAO 2016, *Climate change and food security: risks and responses,* Food and Agriculture Organisation, 2016. http://www.fao.org/3/a-i5188e.pdf

Farzana, FD, Rahman, AS, Sultana, S, Raihan, MJ, Haque, MA, Waid, JL, Choudhury, N & Ahmed, T 2017, 'Coping strategies related to food insecurity at the household level in Bangladesh', *PLoS One* 12:e0171411. doi: 10.1371 journal.pone.0171411.

Gopichandran, V. Claudius, P. Baby, IS. Felinda, A & Mohan, VR 2010, 'Household food security in urban Tamil Nadu: A survey in Vellore, *National Medical Journal of India,* vol.23, pp.278-280.

Gregory, PJ, Ingram, JS & Brklacich, M 2005, 'Climate change and food security', *Philos Trans R Soc Lond B Biol Sci,* vol.360, pp.2139-2148.

Gupta P Singh, K. Seth, V. Agarwal, S & Mathur, P 2015, 'Coping strategies adopted by households to prevent food insecurity in urban slums of Delhi, India', *Journal of Food Security,* vol. 3, pp.6-10.

Islam, S, Mahanta, TG, Sarma, R & Hiranya, H 2014, 'Nutritional status of under 5 children belonging to tribal population living in riverine (Char) areas of Dibrugarh District, Assam', *Indian Journal of Community Medicine,* vol. 39, pp.169-174.

Jones, AD, Ngure, FM, Pelto, G & Young, SL 2013, 'What are we assessing when we measure food security? A compendium rev current metrics', *Advances in Nutrition,* vol.4, pp.481-505.

Joshi, A, Arora, A3, Amadi-Mgbenka, C, Mittal, N, Sharma, S, Malhotra, B, Grover, A, Misra, A & Loomba, M 2019, 'Burden of household food insecurity in urban slum settings', *PLoS One* 14:e0214461. doi: 10.1371/journal.pone.0214461.

Kakoti, N 2017, 'The Mising tribe of North-east India- A glimpse', *International Journal of Management and Applied Science*, vol. 3, no.9, pp.11-13.

Kumar, N, Shekhar, C, Kumar, P & Kundu, AS 2007, 'Kuppuswamy's socio-economic status scale-updating for 2007', *Indian Journal of Pediatrics*, 74:1131-2. doi: t: https://www.researchgate.net/publication/5674708

Longvah, T, Khutsoh, B, Meshram, II, Krishna, S, Kodali, V, Roy, P & Kuhnlein, HV 2017, 'Mother and child nutrition among the Chakhesang tribe in the state of Nagaland, North-east India', *Maternal and Child Nutrition,* Suppl 3. doi: 10.1111/mcn.12558.

Mastiholi, SC, Somannavar, MS, Vernekar, SS, Yogesh Kumar, S, Dhaded, SM, Herekar, VR, Lander, RL, Hambidge, MK, Krebs, NF & Goudar, SS 2018, 'Food insecurity and nutritional status of preconception women in a rural population of North Karnataka, India', *Journal of Reproductive Health,* 2018 Jun 22;15(Suppl 1):90. doi: 10.1186/s12978-018-0535-2.

McDonald, CM, McLean, J, Kroeun, H, Talukder, A, Lynd, LD & Green, TJ 2014, 'Household food insecurity and dietary diversity as correlates of maternal and child undernutrition in rural Cambodia,' *European Journal of Clinical Nutrition,* vol.9, pp. 242-246. doi: 10.1038/ejcn.2014.161.

McDonald, CM, McLean, J, Kroeun, H, Talukder, A, Lynd, LD & Green, TJ 2015, 'Household food insecurity and dietary diversity as correlates of maternal and child undernutrition in rural Cambodia', *European Journal of Clinical Nutrition,* vol.69, no.2, pp.242-246. doi: 10.1038/ejcn.2014.161.

Medecins Sans Frontieres Foundation 2008, *Food Security Assessment Report.* Chittagong Hill Tracts Sajek Union. Bangladesh.

Mishra, D & Singh, HP 2003, 'Kuppuswamy's socio-economic status scale—a revision', *Indian Journal of Pediatrics,* vol.70, pp.273-274.

Moors, E. Singh, T, Siderius, C, Balakrishnan, S & Mishra, A. 2013, 'Climate change and waterborne diarrhoea in northern India: impacts and adaptation strategies', *Science of the Total Environment,* pp.468-469: S139–S151.

Parry, M, Evans, A, Mark, WR & Wheeler, T 2009, *Climate Change and Hunger: Responding to the Challenge.* Rome: World Food Programme. http://www.preventionweb.net/files/12007_wfp212536.pdf

Pasricha, S, Vijaykumar, V, Prashanth, NS, Sudarshan, H, Biggs, BA, Black, J & Shet AA 2009, 'Community-based field research project investigating anaemia amongst young children living in rural Karnataka, India: a cross-sectional study', *BMC Public Health,* vol.9, pp.59.

Portaluppi, F, Smolensky, MH & Touitou, Y 2010, 'Ethics and methods for biological rhythm research on animals and human beings', *Chronobiology International,* vol.27, pp.1911-1929. doi: 10.3109/07420528.2010.516381.

Ramachandran, N 2014, *Persisting Undernutrition in India: Causes, Consequences, and Possible Solutions.* New Delhi: Springer. (https://link.springer.com/book/10.1007%2F978-81-322-1832-6)

Ramakrishna, G & Gaddam SR. D 2014, 'Impact of floods on food security and livelihoods of IDP tribal households: the case of Khammam region of India', *International Journal of Development Economics Sustainability* vol.2, pp.11-24.

Rani, D, Singh, JK, Acharya, D, Paudel, R, Lee, K & Singh, SP 2018, 'Household food insecurity and mental health among teenage girls living in urban slums in Varanasi, India: A cross-sectional study', *International Journal of Environmental Research and Public Health,* 15. pii: E1585. doi: 10.3390/ijerph15081585.

Schmidhuber, J & Francesco, NT 2007, *Global food security under climate change*. PNAS 104:19703–19708

Sikdar, M 2012, 'Prevalence of malnutrition among the Mising children of North-east India: A comparison between four different sets of criteria', *North American Journal of Medical Sciences,* vol.4, pp.305–309.

Taid, TR 1987, *Linguistics of the Tibeto-Burman Area*. Singapore: A short note on Mishing phonology 10, pp.130–137.

Touitou, Y, Portaluppi, F, Smolensky, MH & Rensing, L 2004, 'Ethical principles and standards for the conduct of human and animal biological rhythm research', *Chronobiology International,* vol.21, pp.161-170.doi. https://doi.org/10.1081/CBI-120030045.

White, H & Masset, E 2007, *Assessing Interventions to Improve Child Nutrition: A Theory-Based Impact Evaluation of the Bangladesh Integrated Nutrition Project* y 652:627–652.

World Bank 2010, *Assessing the Impacts of Climate Change and Variability on the Water and Agricultural Sectors and the Policy Implications.*

Chapter - 10

Teenage Pregnancy: An Increased Risk of Prenatal and Infant Mortality Rate among the Aimol Tribe of Manipur

Khangembam Somibabu Meitei

Introduction

An early marriage inevitably puts teenage girls at risk of being pregnant. Teenage pregnancy is one of the immerging problems in post-modern society. Multiple factors are associated with this rising trend in society all over the world. Around 11 percent of the world's teenage pregnancies happen in India (NFHS, 2015-2016). It means that 16 million women between the ages of 15-19 years become mothers each year, one in four Indian women i.e. 26.8 percent are married before 18 according to the latest available 2015-2016 National Family Health Survey (NFHS). It is also one of the major problems in western and many other pacific countries since there are many risks involved with teenage pregnancies, for instance, teenage mothers usually have poor eating habits, lack of parental skills, irresponsible behavior like smoking and drinking, and often have pregnancy complications like premature birth, newborn deaths, anaemic and low birth weight babies, hypertension, hemorrhage, and unsafe abortion, etc. (Makinson, 1985; UNICEF, 2001; Mayor, 2004; WHO, 2012).

Material and methods

The present chapter is based on demographic data collected among the Aimol tribe of Manipur. A door to door survey was conducted for collecting the reproductive profile of the ever-married women. An interview schedule was administered among 305 ever-married women covering thirteen villages of three districts, namely Chandel district, Senapati district, and

Churachandpur district during 2009-2010. Mean and Multivariate analysis were done by using SPSS.

Results

Age at marriage and its distribution is an important factor affecting fertility. It helps to understand the trend of age at marriage in both males and females. The age at marriage is influenced by the socio-cultural and socio-economic factors affecting the demographic facts like population growth and fertility rate.

Table 10.1: Distribution of age at marriage vis-à-vis number of conceptions and live births among Aimol women

Age at marriage	*No. of women*	*No. of conceptions*	*%*	*Mean*	*Live Births*	*Mean*
<14	2	7	0.7%	3.5	7	3.5
15-19	116	439	44.7%	3.78	433	3.73
20-24	117	364	37.1%	3.11	359	3.07
25-29	60	155	15.8%	2.58	150	2.5
30-34	7	14	1.4%	2	11	1.57
35-39	3	3	0.3%	1	2	0.67
Total	305	982	100	3.22	962	3.15

The distribution of age at marriage vis-à-vis the mean number of conceptions and live births among the Aimols reveals that the highest mean number of conceptions (3.78) and the mean number of live birth (3.73) is found in age cohorts 15-19 (Table 10.1). The mean number of conceptions and the mean number of live births steadily decreases with increasing age at marriage, at the same time the variation between the mean number of conceptions and the mean number of live births also relatively increases with increasing age indicating successful live birth decreases with increasing age at marriage. The study shows that younger age at marriage is likely to have a higher fertility rate and gradually decline in fertility with increasing age at marriage.

Aimol women tended to begin childbearing during their teens (Table 10.2). The highest frequency of the 1st pregnancy is found in their early twenties i.e. 20-24 age cohorts (40.65%) which are followed by age cohort 15-19 years (29.84%). The total teenage (13-19 years) pregnancies at 1st conception are 30.17 percent indicating a high percentage of teenage pregnancies. The percentage of 1st pregnancy at age cohort 25-29, 30-34, and 35-39 years are 24.26 percent, 3.28 percent, and 1.64 percent, respectively. In the second pregnancy also the highest percentage is observed

Table 10.2: Age cohort wise pregnancy distribution vis-à-vis number of conceptions among the Aimol women

Age Cohort	1st	%	2nd	%	3rd	%	4th	%	5th	%	6th	%	7th	%	8th	%
<14	1	0.33	0	0.00	0	0.00	0	0.00	0	0.00	0	0.00	0	0.00	0	0.00
15-19	91	29.84	29	11.51	5	2.65	2	1.49	1	1.45	0	0.00	0	0.00	0	0.00
20-24	124	40.65	118	46.82	53	28.04	15	11.19	4	5.80	2	8.00	1	14.28	0	0.00
25-29	74	24.26	77	30.56	84	44.44	54	40.30	17	24.64	5	20.0	3	42.86	0	0.00
30-34	10	3.28	26	10.32	40	21.16	46	34.34	35	50.72	13	52.0	3	42.86	1	100
35-39	5	1.64	2	0.79	7	3.71	15	11.19	10	14.49	4	16.0	0	0.00	0	0.00
>40	0	0.00	0	0.00	0	0.00	2	1.49	2	2.90	1	4.00	0	0.00	0	0.00
Total	305	100	252	100	189	100	134	100	69	100	25	100	7	100	1	100

in the age cohort 20-24 years with 30.56 percent followed by 25-29, 15-19, 30-34, and 35-39 years with 30.56 percent, 11.51 percent, 10.32 percent, and 0.79 percent, respectively. The number of conceptions decreases with the increasing age. Teenage pregnancy and late childbearing up to the age of 40 years are evident in the present study population.

The Age-specific Fertility Rate shows the age-specific rates of live births. It makes it possible to understand the actual reproductive performance of women in different age cohorts. It is one of the important measures of fertility.

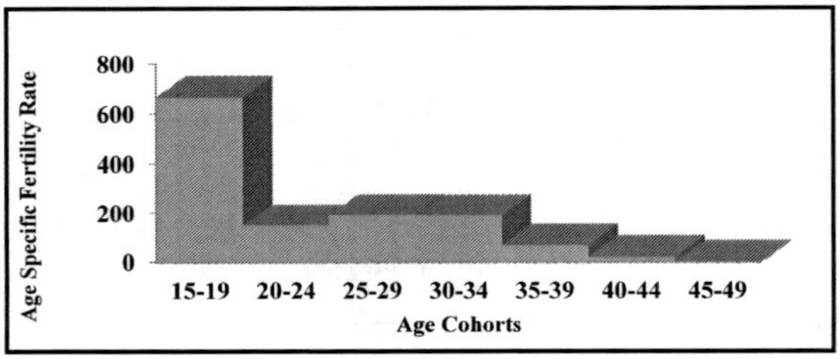

Figure 10.1: Age-Specific Fertility Rate (ASFR) among the Aimols

The age-specific fertility rates among the Aimol reveals that the highest fertility rate is found in the age cohort 15-19 years (666) which is followed by the age cohort 25-29, 30-34, 20-24, 35-39, 40-44 years with 191.49, 188.68, 150, 67.80 and 20.83, respectively (Figure 10.1). It is noteworthy to mention that the peak fertility rate is found in the age cohort 15-19 years indicating that the teenage fertility rate is very high among the Aimol women.

Table 10.3: Stepwise regression analysis of factors influencing fertility among the Aimols

Step Number	*Variable Selected*	*R square*	*P-value*
1.	Current Age of Mother	31.0%	0.000*
2.	Age at Marriage	10.1%	0.000*
3.	Per Capita Income	3.5%	0.000*
4.	Duration of Breastfeeding	1.5%	0.004*
5.	Age at Menarche	0.6%	0.067
6.	Education of the Mother	0.2%	0.264
7.	Age at first conception	0.1%	0.613
8.	Birth control Measures	0.0%	0.676
9.	Post partum Amenorrhea	0.0%	0.751
10	Occupation of Mother	0.0%	0.751

*Statistically Significant.

The stepwise regression analysis of factors influencing fertility rate among the Aimol women reveals that the current age of mother, age at marriage, per capita income, and duration of breastfeeding are found to be statistically significant contributors indicating that these variables are found to be associated with the mean number of conceptions (Table 10.3). Up to 31.0 percent of the variance of the number of conceptions is explained by the current age of the mother, 10.1 percent by age at marriage, 3.5 percent by per capita income, and 1.5 percent by the duration of breastfeeding. Age at menarche, education of mother, and age at first conception variables contributed 0.6 percent, 0.2 percent, and 0.1 percent, respectively though they do not have a significant contribution. Among the Aimol women birth control measures, postpartum amenorrhea and occupation of the mother are found to be a statistically non-significant contribution to fertility.

Table 10.4: Distribution of perinatal, infant, and prenatal mortality vis-à-vis age of mother at childbirth among the Aimol women

Age at conception cohort	*No. of conceptions*	*Prenatal Deaths*	*%*	*Perinatal deaths including Still Births*	*%*	*Infant Mortality (Neonatal, Post neonatal)*	*%*
<14	1	0	0	0	0	0	0
15-19	129	3	33.33	1	8.33	5	41.67
20-24	317	0	0.00	3	25.0	3	25.0
25-29	310	2	22.22	3	25.0	3	25.0
30-34	174	3	33.33	4	33.4	1	8.33
35-39	46	1	11.12	0	0.00	0	0.00
40-44	5	0	0.00	1	8.33	0	0.00
Total	982	9	100	12	100	12	100

The distribution of prenatal, perinatal, and infant mortality rates concerning the age of mother at childbirth reveals that different distribution of mortality rate is found in the different age cohort of mothers (Table 10.4). The prenatal mortality rate and infant mortality rate are high among the teenage cohort 15-19 years with 33.33 percent and 41.67 percent, respectively. The highest Age-Specific Fertility Rate is also observed in this age cohort (Figure 1). The prenatal and perinatal mortality is high among the age cohort 30-34 years indicating the association of prenatal and perinatal risk factors for higher age cohort pregnancies with 33.3 percent and 33.4 percent, respectively. The perinatal mortality rate gradually increases with increasing the age cohort of mothers at childbirth and reverse is the case in the infant mortality rate indicating the reverse association of perinatal death and infant mortality rate with the age cohort of the mother at childbirth.

Perinatal Death and Infant Mortality

It is the number of stillbirths and deaths in the first week of life per thousand live births.

Table 10.5: Stepwise regression analysis of factors influencing perinatal death and infant mortality among the Aimols

Step Number	*Variable selected*	*Perinatal Death*		*Infant Mortality*	
		R square	*P-value*	*R square*	*P-value*
1.	Age at marriage	2.4%	0.106	0.3%	0.462
2.	Prenatal check ups	1.0%	0.100	1.8%	0.061
3.	Age at first conception	0.6%	0.101	0.0%	0.911
4.	Per capita income	0.4%	0.246	0.0%	0.780
5.	Current age of mother	0.3%	0.381	1.6%	0.009*
6.	Education of mother	0.1%	0.626	0.0%	0.888
7.	Occupation	0.0%	0.767	0.2%	0.422
8.	Number of conceptions	2.3%	0.018*		
9.	Antenatal check up	0.2%	0.224		

*Statistically Significant

The stepwise regression analysis of factors influencing perinatal mortality rate among the presently studied population reveals that out of 9 independent variables only 'number of conception' variables is found to be statistically significant (Table 10.5). Up to 2.3 percent of the variance is explained by the number of conceptions in the perinatal mortality rate among the Aimol women. Age at menarche contributes 2.4 percent, prenatal checkups contribute 1.0 percent, age at first conception contributes 0.6 percent, per capita income contributes 0.4 percent, antenatal checkups contribute 0.2 percent, current age of mother contributes 0.3 percent, and education of mother contributes 0.1 percent, though none of them is found to influence the perinatal mortality significantly among Aimol women. The occupation of Aimol women is also found to have a statistically non-significant contribution to perinatal mortality.

Infant mortality is an important parameter of health and development which are influenced by the socio-economic development and quality of life of the population. The stepwise regression analysis of factors influencing infant mortality rate among the Aimol tribe shows that of the seven independent variables only the current age of the mother is found to be statistically significant (Table 10.5). Up to 1.6 percent of the variance is explained by the current age of mothers in the infant mortality rate. Prenatal check-ups, age at marriage, and occupation of women contribute 1.8 percent, 0.3 percent, and 0.2 percent, respectively though none of them influences infant mortality significantly among Aimol women. Per capita income, education of mother, and age at first conception are also found to be a statistically non-significant contribution in infant mortality among the present study population.

Discussion

As per the UNFPA report released in 2013, about 7.3 million girls under the age of 18 give birth to children, India topped the chart of 10 countries during 2000-2013 with the largest number aged between 20 to 24 years who gave birth at the age of 18 (UNFPA, 2013). The greatest threat of teenage pregnancy is the higher rate of pregnancy-related complications like teen mothers usually have poor eating habits, lack parental skills, irresponsible behavior like smoking and drinking, and often have pregnancy complications like premature birth, anaemia, hypertension, hemorrhage, stillbirth, newborn deaths, low birth weight babies, etc. (Markinson, 1985; UNICEF, 2001; Mayor, 2004, WHO report, 2012). WHO report (2012) said that some 70,000 teen girls die every year from complications in pregnancy and childbirth, mostly in developing countries.

Table 10.6: Comparative mean age at the marriage of Manipuri, North-east Indian populations

Population	*Mean age at Marriage*	*Reference*
Aimol	21.1	Present Study
Tribal women of Manipur	21.6	Mukherjee, 1990
Anal women	22.4	Khiloni, 2009
Mao women	22.7	Maheo, 2004
Kabui	20.03	Singh, 2006
Meitei	19.98	Singh, 2006
Assam	23.4	Census, 2013
India	22.5	Census, 2013

The mean age at marriage among the Aimol is 21.1 years which is found to be lower than that of mean tribal women of Manipur (21.6 years), Anal women (22.4 years), Mao (22.7 years), Assam (23.4 years), and India (22.5 years) but higher than that of Meitei (19.98 years) and Kabui (20.03 years) women of Manipur. 82.5 percent of Aimol women get married before 25 years of age (Table 10.1). The mean number of conceptions and mean live births steadily decreases with increasing age at marriage, at the same time the variation between the mean numbers of conceptions and mean live births also relatively increases with increasing age indicating successful live birth decreases with increasing age at marriage. The younger age at marriage seems to have a higher fertility rate and there is gradually a decline in fertility with increasing age at marriage.

The mean age at first conception is 21.75 which is lower than that of Kabui but higher than the Meitei (Singh, 2006). The peak fertility rate (666) is found in the age cohort 15-19 years indicating that the teenage fertility rate is very high among the Aimol women. A higher fertility rate observed in the population could be because of reproductive overcompensation due to higher prenatal and infant mortality rates.

Biological factors like the current age of the mother, age at marriage, and duration of breastfeeding are found to be statistically significant contributors indicating that these variables are found to be associated with the mean number of conceptions. Among the social factors, per capita income is found to contribute to fertility.

The prenatal mortality rate and infant mortality rate are high among the teenage cohort 15-19 years with 33.33 percent and 41.67 percent, respectively. Up to 2.3 percent of the variance is explained by the number of conceptions in the perinatal mortality rate among the Aimol women. Upto1.6 percent of the variance is explained by the current age of the mother in the infant mortality rate.

Conclusion

The Aimol teenage fertility rate is needed to be checked to reduce fertility, prenatal, and infant mortality rate among teenage cohorts. Teenage marriages, education of women, and knowledge of factors affecting infant mortality is a fundamental requirement for devising appropriate policies and strategies for reducing the infant mortality rate among the present studied population. Effective interventions for teen pregnancy prevention and appropriate reproductive healthcare services, awareness of maternal and child health care programs are needed for young mothers and families to reduce the infant mortality rate of the study population. The population planners and the family welfare programmers could make use of the present study in checking the fertility status and infant mortality rate of the Aimol community. It is high time to bridge the gap and addressing the sensitive social and medical issues associated with teenage pregnancy.

References

International Institute for Population Sciences (IIPS) and Macro International. *National Family Health Survey 4,* 2015-16: India Fact sheet. Mumbai; 2017.

Khiloni, L 2009, Fertility performance of the Anal women of Lambung village, Chandel District, Manipur, *Anthropologist*, vol.11, no.4, pp.277-280.

Maheo, LM 2004, *The Mao Naga Tribe of Manipur-A Demographic Anthropological Study*. New Delhi: Mittal Publications.

Makinson, C 1985, The health consequences of teenage fertility, *Family Planning Perspectives*, vol.17, pp.132-139.

Mayor, S 2004, Pregnancy and childbirth are leading causes of death in teenage girls in developing countries, *British Medical Journal*, vol.328, no.7449, p.1152.

Mukherjee, BM 1990, Health status of Manipuri women with special reference to the tribals, in Chaudhuri, B. (ed.) *Cultural and Environmental Dimensions on Health*. Delhi: Inter-India Publications. pp. 80-100.

Singh, SJ 2006, Ethnic variation in fertility patterns among four communities of Manipur, *Journal of Human Ecology*, vol.20, no.1, pp.1-9.

UNICEF 2001, *A league table of teenage births in rich nations*. Innocenti Report Card No. 3. Florence: UNICEF Innocenti Research Centre.

WHO Fact sheet No.364 May 2012, http://www.unfpa.org/adolescent-pregnancy

Chapter - 11

Maternal Genetic Structure among the Manipuri Population Reflected through Mitochondria DNA Sequence

S.Y. Meitei, K.S. Meitei, N. Achoubi Devi, M.P. Sachdeva, K.N.Saraswathy

Introduction

Manipur is one of the small hilly states that lie on the extreme North-eastern border of India. It has been the homeland for various ethnic groups, including both tribal and non-tribal communities, since prehistoric times. Because of her strategic geographical situation on the Indo-Myanmar border, the state has been considered as one of the gates through which Mongoloid tribes, their languages, and beliefs poured into eastern India (Chakravartti, 1986). Though the state has been geographically isolated from mainland India, the migratory movements of Aryans from the west and Mongoloid people from south-east-Asia were evident through the terrain of the valley (Ali and Das, 2003). It seems that Mongoloid stock from East and South-East Asia have mixed with the Proto-Australoid and Indo-Aryan strains from mainland India. Manipur is the place where different waves of races and cultures met through the ages, which ultimately mingled together (Zehol, 1998). The North-east tribes including the Manipur ones are believed to have migrated from East Asia and South East Asia through two possible routes – Tibet and Yunan province of China (Ansari, 1991). Meitei, a non-tribal population, is the largest ethnic group in the state and claimed to be original settlers. Though the history of Meitei is not very clear, Hodson (1908) suggested that Meiteis are presumably formed by the admixture of Koomal, Looang, Moirang, and Meitei, all of whom came in different periods from different directions; and now representing the clans of the community. It has also been hypothesized that the Meitei is the product of the immigrated Mongoloid elements and the already existing proto Australoid element of the region (Kabui, 1989) and genetic studies support this hypothesis (Saraswathy *et al.,* 2009a,b).

In the present study, we analyzed the mtDNA control region hypervariable segment I sequence (HVSI) in four ethnic population groups of Manipur and tried to understand the possible relationship of these population ethnohistories with their genetic architecture through nucleotide and haplotypic diversity among these populations. The studied populations include both the tribal – Paite, Kom, Thadou, and non-tribal communities (Meitei) of Manipur and all of them belong to the same racial stock – Mongoloid. Meitei speak Naga-Bodo groups of Tibeto-Burman linguistic family while all the three tribal communities speak the Kuki-Chin group of the same linguistic family

Materials and methods

Genomic DNA was extracted from the intravenous blood samples collected from each population group, after prior informed consent. The control region of mtDNA HVSI of 73 unrelated individuals that belonged to the mentioned population groups: Meitei (n = 16), Kom (n = 19), Paite (n = 18), and Thadou (n=20) were amplified using primer MT HV SIF 5′-CTC CAC CAT TAG CAC CCA AAG-3′ and MT HV SIR 5′-TGA TTT CAC GGA GGA TGG TG-3′. The amplified product was sequenced in TCGA, Delhi. The sequenced data were aligned using CLUSTAL W and checked with the CRS sequence (Anderson *et al.,* 1981). The analysis is restricted to 357bp length only (nucleotide position 16024 -16380). The number of shared haplotypes among the studied population groups was identified. Inter and intrapopulation diversity among these population groups was analyzed using statistical software – Arlequin, DnaSP. The mismatch analysis under the population growth and decline was analyzed and the graphical representation of the analysis was made using DnaSP. Assuming the rate of divergence of the mtDNA control region as 33.0 percent site/ million years (Ward *et al.,* 1991), the approximate year of population expansion was calculated using the formula, ô = 2tì. The genetic structure among these populations was studied through AMOVA. An NJ tree was drawn using the net pairwise genetic distance and the corresponding consensus tree was also drawn using the software Phylip 3.69.

Results

The sequence data of mtDNA HVSI are found to be highly polymorphic in the presently studied populations. Out of the studied populations, Thadou exhibits the highest polymorphic at 184 nucleotide positions and the least among the Kom tribe at 30 nucleotide sites (Table 11.1). The maximum number of haplotypes is also found in Thadou (19 haplotypes) and the least among the Kom (11 haplotypes). Both Meitei and Paite have the same number of haplotypes (15 each). Both the Paite and the Thadou share the same two haplotypes thereby resulting in 86.66 percent and 89.47 percent unique haplotypes, respectively, while all haplotypes are unique in both Meitei and Kom.

Table 11.1: Sample information of populations analyzed in the present study

Population (n)	*Ethnicity (Census size)**	*Expected origin*	*No. of polymorphic sites*	*Expected heterozygosity*	*No. of haplotypes*	*No. of unique haplotype (%)*
Meitei (16)	~1,433,180	China (Pemberton, 1835)	57	0.026	15	15 (100%)
Kom (19)	~15,467	Myanmar (Southeast Asia) (Shakespear, 1912)	30	0.021	11	11 (100%)
Thadou (20)	~115,045	Myanmar (Southeast Asia) (Shakespear, 1912)	184	0.098	19	17 (89.47%)
Paite (18)	~44,861	Myanmar (Southeast Asia) (Shakespear, 1912)	116	0.048	15	13 (86.66%)

*Census, 2011

Table 11.2: Genetic diversity and population demographic parameters in the population

Population (n)	*Mean no. of pairwise differences*	*Gene diversity*	*Haplotype diversity*	*Nucleotide diversity*	*D*	*Tau*	*Demographic expansion time (YBP)*
Meitei (16)	10.092 ± 4.87	0.992 ± 0.03	0.9917	0.028 ± 0.02	-1.719*	4.986	41,900
Kom (19)	8.129 ± 3.95	0.912 ± 0.04	0.9123	0.023 ± 0.01	-0.003	9.969	83,900
Thadou (20)	37.674 ± 17.10	0.995 ± 0.02	0.9947	0.104 ± 0.05	-1.235	5.988	50,400
Paite (18)	18.327 ± 8.5	0.961 ± 0.04	0.9608	0.050 ± 0.03	-1.910*	7.383	62,100

*significant at 0.05

Regarding the intra-populational diversity among the presently studied populations (Table 11.2), gene diversity ranges from 0.912 (Kom) to 0.995 (Thadou). Thadou exhibits the highest haplotype and nucleotide diversity among all the studied populations while Kom has the least. The mean number of pairwise differences among these populations is following the

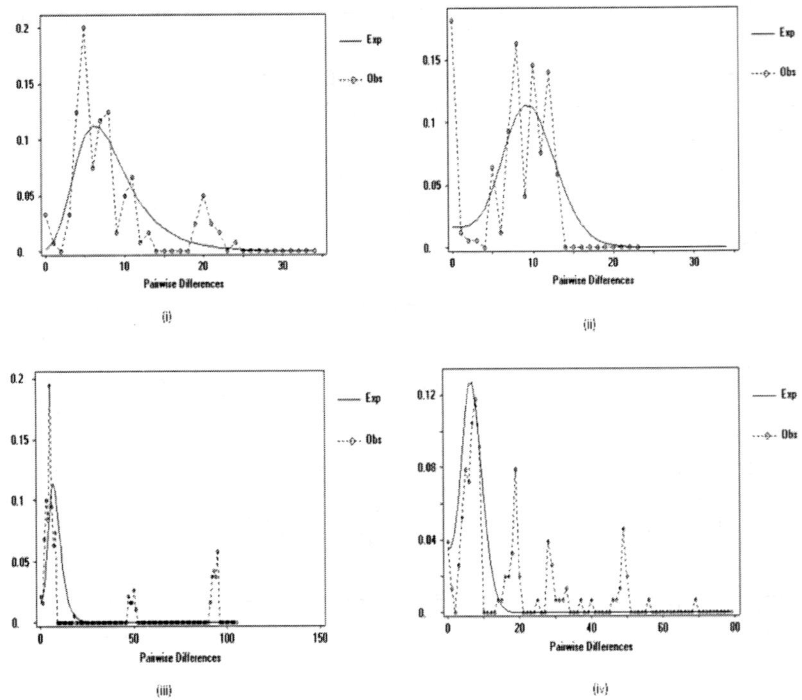

Figure 11.1: Mismatch distribution between the pairwise differences of mtDNA nucleotides i) Meitei, ii) Paite, iii) Kom and iv) Thadou

number of polymorphic sites, i.e. the population group exhibiting the highest polymorphic sites has the maximum mean number of pairwise differences. Tajima D value ranges from -0.003 to -1.910 which is statistically significant among Meitei ($p = 0.022$, $p < 0.05$) and Paite ($p = 0.015$, $p < 0.05$). Demographic expansion time calculated from the tau value (ô) of each population places them around 62,100 YBP expansion except Kom which is 83,900 YBP expansion. The mismatch distribution between the pairwise differences of nucleotides reveals multimodal distribution, except Thadou which exhibits ragged distribution (Figure 11.1).

Discussion

The number of unique haplotypes is quite high reaching almost 100 percent as compared with that of the East and South-east Asian population where the unique haplotypes range from 44.4 percent to 90.0 percent. This highly polymorphic nature of mtDNA HVSI in the presently studied population is concordance with the earlier studies using classical and molecular markers (Roychudhury *et al.,* 2001; Dutta *et al.,* 2003; Saraswathy *et al.,* 2009 a,b) where these populations are reported to be highly heterogeneous. These populations including the whole of North-east India are expected to be homogeneous as they belong to the same Mongoloid stock though they speak different dialects under the same Tibeto-Buman linguistic family, except Khasi who speaks Mon-Khmer group of Austro-Asiatic linguistic family. Moreover, these tribes have almost similar cultural practices. The sharing of two haplotypes among Paite and Thadou supports Saraswathy *et al.,* (2009a)'s proposal of the genetic proximity of these two tribes, which further indicates the commonality in the maternal gene pool of these populations, as mtDNA is maternally inherited.

Gene diversity which is defined as the probability that two randomly chosen haplotypes are different in the sample is equivalent to the expected heterozygosity for diploid data. The nucleotide (ð) and haplotype diversity (h) values (0.023 ± 0.02 to 0.104 ± 0.05; and 0.9123 to 0.9947, respectively) are almost same with that of Chinese populations. Though all the intrapopulation diversity indexes are almost in uniform, the Kom tribe exhibits the least in all the indexes which could be explained due to small census population sizes. This could be also possible for the severe population size reduction in the past.

To understand the selective neutrality among the presently studied populations, Tajima's D was calculated, and it is found to vary from -0.003 to -1.91. This value is found to be statistically significant among the Meitei ($p = 0.022$) and the Paite (0.015). This significant D value among Meitei is expected as it is an admixed group, while that among Paite could be explained due to bottleneck (Saraswathy *et al.,* 2009a). The tau value (ô), which reflects the location of the mismatch distribution crest, provides a rough estimate of the time when rapid population expansion started (Rogers and Harpending, 1992; Rogers, 1995). Assuming the divergent rate of the mtDNA control region to be 33.0 percent site/million years (Ward *et al.,* 1991), the approximate ages of population expansion of these groups were calculated. These populations have an expansion period of more than ~42,000 YBP (42,000 – 62,000 YBP), which is in concordance with the age of population expansion of the East Asian population (~ 33,100 – 68,300; Yao *et al.,* 2002). This finding is consistent with the possible year of initial Palaeolithic expansion in East Asia as suggested by mtDNA HVSI sequence analysis (Yao *et al.,* 2002) and Y chromosome biallelic markers (Su *et al.,* 1999). Kom tribe, which is the smallest population group and less polymorphic in the present study, has a very early age of expansion of around 83,000 YBP, which needs to be explained. The previous signal of Pleistocene expansion might have been erased in this population due to episodes of small sample size (Excoffier and Schneider, 1999). The multimodal distribution in the mismatch analysis reveals that the populations under study are in equilibrium (Rogers and Harpending, 1992). Since the census sizes of Meitei, Thadou, and Paite are large and the samplings of the populations were restricted to only a few villages of these ethnic populations, the ragged mismatch distributions observed might be restricted to these sampled population analyzed only, not entirely in the population group. Moreover, the mismatch distributions could be easily affected by the sample sizes considered (Yao *et al.,* 2002).

Table 11.3: AMOVA analysis to understand the genetic structure

Genetic structure	*% of variation*	*P value*
Among populations	3.74	0.004 ± 0.002
Within populations	96.26	

The genetic structure of the presently studied ethnic groups was investigated by AMOVA (Table 11.3; Excoffer *et al.,* 1992). Considering haplotypes, 96.26 percent of the genetic variation was found within population groups, whereas 3.74 percent of the variation ($p<0.05$) was among the studied population groups. The significant inter-population genetic variation confirms the highly heterogeneous of the presently studied population, which further hints to different maternal genetic contribution to these population groups

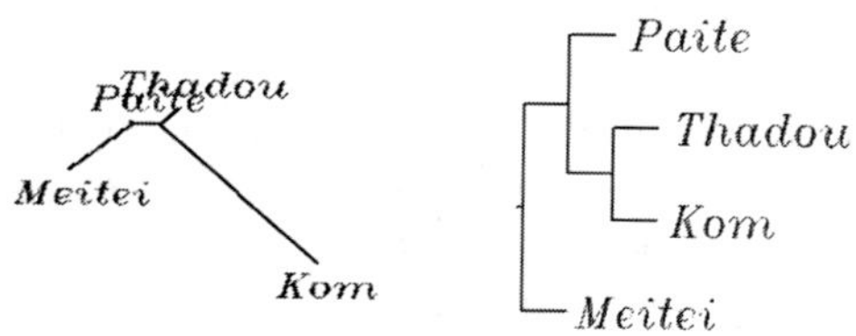

(i) (ii)

Figure 11.2: Neighbour-Joining Tree depicting population relationship

Unrooted Neighbour-Joining Tree (Figure 11.2) of the presently studied populations was constructed using the net pairwise genetic distance using Phylip 3.69. In the tree, Paite and Thadou lie very close to each other which further supports the genetic proximity of these two tribes. Kom tribe lies separately from all the studied populations. The relatively large genetic distances between Kom and other populations could also result from the impact of genetic drift in this small tribe (Relethford, 1996; Excoffier and Schneider, 1999). Yet, there is a commonality in the consensus tree drawn using the net pairwise genetic distance (Figure 11.2ii), that distinguishes all the three tribal populations from the non-tribal population i.e. Meitei. All the three tribal populations could be traced back to a common ancestry line in the tree. These tribal populations are categorized under the Kuki tribes. Based on the time frame of the migration, linguistic and cultural similarities, the Kuki tribes are further classified into two tribal groups – Old Kuki and New Kuki (Shakespear, 1912). Being Kom belongs to Old Kuki and Paite and Thadou belong to New Kuki, which supports the genetic differentiation between Kom on one hand and Paite and Thadou on the other hand. The common ancestry branch in these three tribes as observed in the consensus tree could be explained with their common tribal group – Kuki, despite Old and New Kuki. It could be further hypothesized that these tribes might have common maternal genetic constituents a long time back, which had been diverged due to their different waves of migration and social isolation of these tribes.

Conclusion

The Mongoloid tribes inhabiting North-east India including Manipur are believed that they entered the present geographical region through two possible routes – through Tibet and the Yunnan province of China (Ansari, 1991). Those that have migrated through Tibet side came across the Himalayas and settled in Arunachal Pradesh, Brahmaputra Valley of Assam, Nagaland, north-western Burma, and almost the Northern half of Manipur. They speak different dialects belonging to the Naga group of Tibeto-Burman linguistic family. And those following the routes through the Yunnan and eastern China moved southwards to the Islands of South East Asia and Burma, and further westward to the hill region of Indo-Burma frontier tract and subsequently to the Northern half of Manipur too. They speak different dialects belonging to the Kuki-Chin group of the Tibeto-Burman linguistic family. The immigration of certain tribes especially falls under the Kuki tribe into Manipur, in the later age of the 20th century has been reported.

Different waves of migration of smaller groups at different times may lead to a large number of ethnic communities in the state which become more rigid because of strict tribe endogamy and social isolation. This further may lead to the highly polymorphic nature of the presently studied populations which is confirmed by AMOVA analysis. Less diversity among

Kom reveals the severe reduction of the population sizes in the past. Sharing of haplotypes and closely placed in NJ tree among Paite and Thadou assure that there is genetic proximity among them with the possibility of sharing the maternal gene pool. The commonness of ancestral line among three tribal groups supports the mythologically claimed of being the Kuki tribe, which further hypothesized that the present Kuki tribal groups might have separated and distributed in different areas and at different times from a single ancestral community. Admixture might be the reason for the Meitei population being distantly placed from the other studied tribal groups, which is further evident in Tajima's D test. In conclusion, it can be inferred that the presently studied population of Manipur might have reached the state during the early prehistoric expansion of East Asia, and differentiate out among them because of their long ethnohistory followed by strict tribal endogamy. Although, for a better understanding of ancient Paleolithic migration in the present area, a multi populational study covering East Asia and the whole North-east Indian population would be the ideal approach.

References

Ali, ANM Irshad & Das, I .2003, 'Tribal situation in North-east India', *Studies of Tribes Tribals*, vol.1, no.2, pp.141-148.

Anderson, S, Bankier, AT, Barrellm, BG, De Bruijn, HL, Coulson, AR, Drouin, J, Eperon, IC, Nierlich, DP, Roe, BA, Sanger, F, Schreier, PH, Smith, AJH, Staden, R & Young, IG 1981, 'Sequence and organization of the human mitochondrial genome', *Nature,* vol. 290, pp.457-465.

Ansari, SA 1991, *Manipur: Tribal Demography and Socio-Economic Development*. Delhi: Daya Publishing House.

Chakravartti, R 1986, *People of Manipur: Anthropogenetic study of four Manipur population groups*. Delhi: B.R. Publishing.

Dutta, R, Kashyap, VK, Trivedi, R, & Vasulu, TS 2003, 'Genomic diversity in North-east India: Genetic relationship among five ethnic populations from Manipur based on VNTR Polymorphism', *International Journal of Human Genetics,* vol.3, no.1, pp.21-27.

Excoffier, L & Schneider, S 1999, 'Why hunter-gatherer populations do not show sign of Pleistocene demographic expansions,' *Proceedings of the National Academy of Sciences of USA,* vol.96, pp.10597-10602.

Excoffier, L, Smouse, PE & Quattro, JM 1992, 'Analysis of molecular variance inferred from metric distances among DNA haplotypes: application to human mitochondrial DNA restriction data', *Genetics,* vol.131, pp.479–491.

Hodson, TC 1908, *The Meitheis*. Delhi: Low price publication.

Kabui, G 1989, *History of Manipur*. Vol 1. New Delhi: National Publishing House.

Relethford, JH 1996, 'Genetic drift can obscure population history: Problem and solution', *Human Biology*, vol.68, pp.29-44.

Rogers, AR 1995, 'Genetic evidence for a Pleistocene population expansion', *Evolution,* vol.49, pp.608-615.

Rogers, AR & Harpending, H 1992, 'Population growth makes waves in the distribution of pairwise genetic differences', *Molecular Biology and Evolution,* vol.9, pp.552-569.

Roychoudhury, S, Roy, S, Basu, A, Banerjee, R, Vishwanathan, H, Usha Rani, MV, Sil, SK, Mitra, M & Majumder, PP 2001, 'Genomic structures and population histories of linguistically distinct tribal groups of India', *Journal of Human Genetics,* vol.109, pp.339–350.

Saraswathy KN, Kiranmala, N, Murry, B, Sinha, E, Saksena, D, Kaur, H, Sachdeva, MP & Kalla, AK 2009a, 'A genomic insight into diversity among Tribal and Non-tribal population groups of Manipur, India', *Biochemical Genetics*, vol.47, pp. 694-706.

Saraswathy, KN, Meitei, SY, Murry, B, Devi, NK, Sinha, E, Gupta, V, Mukhopadhyay, R, Asghar, Md, Devi, NA, Meitei, KS & Sachdeva, MP 2009b, 'Population severance in Manipur at DRD2 locus', *Genetic Testing and Molecular Biomarkers,* vol.13, pp.831–839.

Shakespear, LW 1912, *The Lushai Kuki Clans*. London: MacMillan.

Su, B, Xiao, JH, Underhill, P, Deka, R, Zhang, WL, Akey, J, Huang, W, Shen, D, Lu, D, Luo, J, Chu, J, Tan, J, Shen, P, Davis, R, Cavalli-Sforza, L, Chakraborty, R, Xiong, M, Du, R, Oefner, P, Chen, Z & Jin, L 1999, 'Y-Chromosome evidence for a northward migration of modern humans into Eastern Asia during the last Ice Age', *American Journal of Human Genetics,* vol.65, no.6, pp.1718-1724.

Ward, RH, Frazier, BL, Dew-Jager, K & Paabo, S 1991, 'Extensive mitochondrial diversity within a single American tribe', *Proceedings of the National Academy of Sciences of USA,* vol.88, pp.8720-8724.

Yao, YG, Kong, QP, Bandelt, HJ, Kivisild, T & Zhang, YP 2002, 'Phylogeographic differentiation of mitochondrial DNA in Han Chinese', *American Journal of Human Genetics,* vol.70, pp.635–651.

Zehol, L 1998, *Ethnicity in Manipur: Experiences issues and perspectives*. New Delhi: Regency Publications.

Chapter - 12

ACE Gene Polymorphism among the Rongmei and Inpui Tribes of Manipur

GANGAINA KAMEIH AND GAUTAM K. KSHATRIYA

Introduction

The Angiotensin I-converting enzyme (ACE) gene is a candidate gene involved in the renin-angiotensin-aldosterone system (RAAS). The ACE enzyme converts inactive angiotensin I into angiotensin II and also degrades bradykinin to maintain homeostasis of blood pressure (Baudin 2002; Coates 2003; Fleming 2006). The ACE insertion/deletion (I/D) gene polymorphism of a 287 Alu element is located in intron 16 on chromosome 17. Despite its location in the non-coding region, ACE I/D gene polymorphism appears to regulate the serum ACE activity.

Several epidemiological have reported a positive association of ACE gene I/D with heart disease, hypertension, diabetes, and diabetic nephropathy (Doi *et al.,* 1996, Ohno *et al.,* 1996, Yoshida *et al.,* 1996, Staessen *et al.,* 1997 and Kennon *et al.,* 1999). ACE is also one of the candidate genes for various complex disorders like IgA Nephropathy (Yoshida *et al.,* 1995), Meningococcal disease (Harding *et al.,* 2002), Preterm Cardio-respiration (Harding *et al.,* 2003), Alzheimer disease (Elkins *et al.,* 2004), and Severe Acute Respiratory Syndrome (Itoyama *et al.,* 2004).

Polymorphism of the ACE gene also varies from population to population. Hence, it is important to have baseline data on the frequency distribution of known and clinically important mutations of the particular gene among various populations especially the tribal population of Manipur which is under-reported for both clinical and genetic data. Therefore, in the present study, an attempt is made to understand the extent of ACE gene polymorphism among the Rongmei and Inpui tribes of Manipur.

The total population of Rongmei and Inpui are 90,372 (Tohring, 2010) and 11,000 (Devi, 2013) respectively. Rongmei and Inpui are found in Manipur, Assam, and Nagaland. About their area of distribution, they mainly concentrate on Manipur in Tamenglong district and along with certain areas of Tadubi Sub-division of Senapati district, Loktak Project region of Churachandpur, Bishnupur and Imphal valley. In Nagaland, they concentrate in Peren district and few pockets along Kohima and Dimapur. Also, they occupy some areas in Assam like North-east of North Cachar Hills district and Cachar valley of Assam. The tribes of Naga (including Rongmei and Inpui) are believed to have migrated from South-west China (Kamei 2015; Kameih *et al.,* 2018).

Material and methods

ACE gene polymorphism was analyzed among 210 individuals belonging to the Rongmei and the Inpui tribes of Manipur. The details of the study populations are given in table 1. Five milliliters of the intravenous blood sample was collected from unrelated individuals by a trained medical practitioner after taking written consent from the subjects. The ethical clearance was obtained from the Ethical Committee of the Department of Anthropology, University of Delhi.

Table 12.1: Name of the study population groups with sample sizes and area of sample collection

Population	*Sample Size*	*Area of sample collection*
Rongmei	110	Khoupum (Tamenglong District) and Tarung (Imphal West District)
Inpui	100	Haochong (Tamenglong District)

DNA extraction was carried out using the salting-out method (Miller *et al.,* 1988). Quantification was done by a Nanodrop spectrophotometer. The ACE gene mutation was detected by the polymerase chain reaction (PCR) amplification analysis (Majumder *et al.,* 1999), and fragments of about 190 bp (D allele) and 490 bp (I allele) were separated on 2 percent agarose gel electrophoresis and the genotypes were visualized with long-wave UV light with ethidium bromide as a staining agent. Allele frequency was obtained using the software POPGENE version 1.31. (Yeh and Yang, 1999). Hardy Weinberg equilibrium was calculated using the test for the chi-square goodness of fit. The distribution pattern of D allele frequency was depicted using MS Excel 2010.

Results and discussion

Table 12.2 presents data about genotype and allele frequency among the Rongmei and the Inpui at the ACE locus. This locus was found to be polymorphic. A higher frequency of genotype heterozygous (ID) was found in both Rongmei and Inpui. The allele frequency distribution showed a higher frequency of insertion allele than the deletion allele. When applying the chi-square goodness of fit test to determine whether the phenotype and genotype frequencies depart from Hardy-Weinberg proportions and using Bonferroni's correction, it is observed that the phenotype (genotype) frequencies for this locus is in reasonable agreement with their respective Hardy-Weinberg expectations and do not show a significant departure from expected frequencies.

Table 12.2: Genotype and allele frequency distribution at ACE locus among the Rongmei and the Inpui tribes

Population	*Genotype Frequency*			*Allele Frequency*		χ^2	*P-value*
	II	*ID*	*DD*	*I*	*D*		
Rongmei	29	45	27	0.510	0.490	1.302	0.254
Inpui	28	45	27	0.505	0.495	1.101	0.294

I Presence of restriction site; D Absent of restriction site; χ^2-values are non-significant at df=1 and p>0.05

The frequency of the D allele of ACE ID polymorphism is higher among the Inpui (49.5%) as compared to that of the Rongmei (49.0%). However, the difference between the two tribal groups is statistically not significant concerning the genotype distribution.

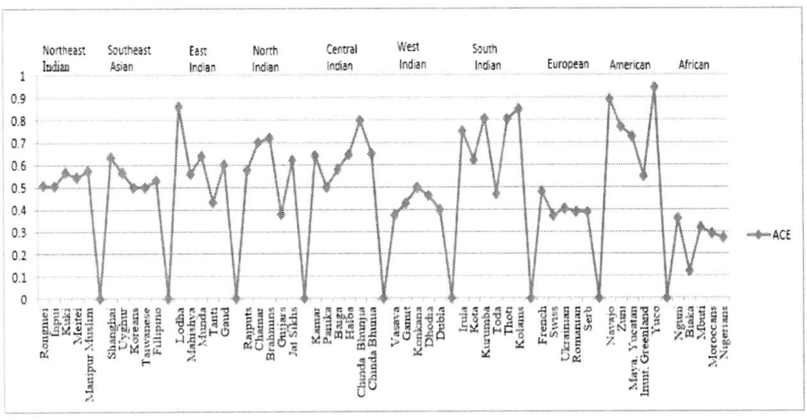

Figure 12.1: Distribution of D allele frequency among Indian and the World populations.

The distribution of D allele frequency of the present studied populations is comparable to the South-east Asian populations but it is found to vary with the other Indian populations (Mukherjee *et al.*, 2000; Vishwanathan *et al.*, 2003; Meitei *et al.*, 2010; Kshatriya *et al.*, 2011; Panjaliya *et al.*, 2012; Gangaina and Tabitha 2015) and the global populations (Xiao *et al.*, 2002; ALFRED database).

Acknowledgement

The help rendered by Dr. Tabitha Panmei and the people of Rongmei, Inpui, Thadou, and Vaiphei during the collection of blood samples is highly appreciated. We express our gratitude to Dr. Sanju Kumar, Department of Biotechnology, Manipur University, for his help by providing laboratory facilities for DNA extraction. We also acknowledge the University Grant Commission for providing financial assistance through JRF.

References

Baudin, B 2002, 'New aspects on angiotensin-converting enzyme: from gene to disease', *Clinical chemistry and laboratory medicine*, vol.40, pp. 256-265.

Coates, D 2003, 'The angiotensin-converting enzyme (ACE)', *The international journal of biochemistry & cell biology,* vol.35, pp.769-773.

Devi, WP 2013, 'Inpui Adverbs', *International Journal of English and Education*, Vol.2, pp. 315-323.

Doi, Y, Yoshzimi, H, Yoshinari, M, Lion, K, Yamamoto, M, Ichikawa, K, Iwase, M & Fujisima, M 1996, 'Association between a polymorphism in the angiotensin-converting enzyme gene and microvascular complication in Japanese patients with NIDDM', *Diabetologia,* vol.39, pp.97-102.

Elkins, JS, Douglas, VC & Johnston, SC 2004, 'Alzheimer disease risk and genetic variation in ACE A meta-analysis', *Neurology,* vol.62, pp.363-368.

Fleming, I 2006, 'Signaling by the angiotensin-converting enzyme', *Circulation Research,* vol. 98, pp.887-896.

Harding, D, Baines, PB, Brull, D, Vassiliou, V, Ellis, I, Hart, A & Montgomery, HE 2002, 'Severity of meningococcal disease in children and the angiotensin-converting enzyme insertion/deletion polymorphism', *American journal of respiratory and critical care medicine,* vol.165, pp.1103-1106.

Harding, D, Dhamrait, S, Marlow, N, Whitelaw, A, Gupta, S, Humphries, S & Montgomery, H 2003, 'Angiotensin-converting enzyme DD genotype is associated with worse perinatal cardiorespiratory adaptation in preterm infants', *The Journal of pediatrics,* vol.143, pp.746-749.

Itoyama, S, Keicho, N, Quy, T, Phi, NC, Long, HT, Van Ban, V & Sasazuki, T 2004, 'ACE1 polymorphism and progression of SARS', *Biochemical and biophysical research communications,* vol.323, pp.1124-1129.

Kamei, B 2015, *Origin, migration, and settlement: The Zeliangrong of NE-Part 4*, The Sangai Express. April 26. Available online at http://e-pao.net/epSubPageExtractor.asp?src=manipur.Ethnic_Races_Manipur.Origin_migration_and_settlement_The_Zeliangrong_of_NE_Part_4_By_Budha_Kamei, accessed January 2018.

Kameih, G, Panmei, T & Kshatriya, GK 2018, 'Allelic variation and haplotype diversity of the Dopamine receptor gene DRD2 among the four sub-populations of Zeliangrong Naga of Manipur', *Journal of Indian Anthropological Society,* vol. 58, pp. 249-259.

Kameih, G & Panmei, T 2015, 'ACE gene polymorphism among Panika and Baiga of Madhya Pradesh', *Human Biology Review,* vol. 4, pp. 303-311.

Kennon, B, Petrie, JR, Small, M & Connell JMC 1999, 'Angiotensin-converting enzyme gene and diabetes mellitus', *Diabetic Medicine,* vol. 16, pp. 448-458.

Kshatriya, GK, Aggarwal, A, Khurana, P & Italia, YM 2011, 'Genomic congruence of Indo-European speaking tribes of western India with Dravidian-speaking populations of southern India: A study of 20 autosomal DNA markers', *Annals of Human Biology,* vol. 38, pp. 583-591.

Majumder, PP, Roy, B, Banerjee, S, Chakraborty, M, Dey, B, Mukherjee, N, Monami, R, Piyali, GT & Samir, KS 1999, 'Human-specific insertion/deletion polymorphisms in Indian populations and their possible evolutionary implications'. *European Journal of Human Genetics,* vol. 7, pp. 435-446.

Meitei, KS, Meitei, SY, Asghar, M, Achoubi, N, Murry, B, Mondal, PR, Sachdeva, MP & Saraswathy, KN 2010, 'A genomic insight into the peopling of Manipur, India', *Genetic Testing and Molecular Biomarkers,* vol.14, pp. 65-773.

Miller, SA, Dykes, DD & Polysky, HF 1988, 'A simple salting out procedure for extracting DNA from human nucleated cells'. *Nucleic Acids Research,* 16: 1215.

Mukherjee, N, Mitra, M, Chakraborty, M & Majumder, PP 2000, 'Congruence of genomic and ethnolinguistic affinities among five tribal populations of Madhya Pradesh (India)', *Journal of Genetics,* vol. 79, pp. 41-46.

Ohno, T, Kawazu, S & Tomono, S 1996, 'Association analysis of the polymorphisms of angiotensin-converting enzyme and angiotensinogen genes with diabetic nephropathy in Japanese non-insulin-dependent diabetics', *Metabolism,* vol. 45, pp. 218-222.

Panjaliya, RK, Dogra, V, Kumar, P & Gupta, S 2012, 'Human-specific Alu Insertion/deletion polymorphisms in various population groups of Jammu region', *International Journal of Human Genetics,* vol. 12, pp. 311-317.

Staessen, JA, Wang, JG, Ginochio, G, Petrov, V, Saavedra, AP, Soubrier, F, Vlietinck, R & Fagard, R 1997, 'The deletion/insertion polymorphism of the angiotensin-converting enzyme gene and cardiovascular-renal risk', *Journal of hypertension,* vol. 15, pp. 1579-1592.

Tohring, SR 2010, *Violence and identity in North-east India: Naga-Kuki conflict*. Mittal Publications. New Delhi.

Vishwanathan, ED, Usharani, MV & Majumder, PP 2003, 'Insertion/deletion polymorphisms in tribal populations of southern India and their possible evolutionary implications,' *Human Biology,* vol. 75, pp. 873-887.

Xiao, F, Yang, J, Cassiman, J & Decorte, R 2002, 'Diversity at eight polymorphic Alu insertion loci in Chinese populations show evidence for European admixture in an ethnic minority population from North-west China', *Human Biology* 74: pp. 555-568.

Yeh, F & Yang, R 1999, *POPGENE. Microsoft Window-Based Freeware for Population Genetic Analysis.* Version 1. 31. University of Alberta, Edmonton, Alberta, Canada.

Yoshida, H, Kuriyama, S, Atsumi, Y, Tomonari, H, Mitarai, T, Hamaguchi, A, Kubo, H, Kawaguchi, Y, Kon, V, Matsuoka, K, Ichikawa, I & Sakai, O 1996, 'Angiotensin I converting enzyme gene polymorphism in non-insulin-dependent diabetes mellitus', *Kidney International,* vol. 50, pp. 657-664.

Yoshida, H, Mitarai, T, Kawamura, T, Kitajima, T, Miyazaki, Y, Nagasawa, R & Sakai, O 1995, 'Role of the deletion of polymorphism of the angiotensin-converting enzyme gene in the progression and therapeutic responsiveness of IgA nephropathy', *Journal of Clinical Investigation,* vol. 9: 2162.

Chapter - 13

Indigenous Health Care Practices: A Case Study of Hilghat Village, Manipur

Malvika Yumnam and Huidrom Suraj Singh

Introduction

Indigenous people of different cultures have well-established diverse health care practices to live well and to free from any unwanted health conditions. Traditional medicine according to World Health Organization (WHO) is described as the knowledge, skills, and practices based on the theories, beliefs, and experiences indigenous to different cultures, used in the maintenance of health and the prevention, diagnosis, improvement, or treatment of physical and mental illness (WHO 2001).

The traditional medicine system examines and translates health-related knowledge that people inherit and learn by living in a culture. It has interrelated notions about the body, the causes, and prevention of illness, diagnosis, and treatment. Traditional medicine examines indigenous ideas and behaviors surrounding how to stay well and treat sickness. Further, it also seeks to understand the internal logic of medical systems. Ethnomedicine is the area of anthropology that studies different societies' notions of health and illness, including how people think and how people act about well being and healing of any illness (Foster and Anderson 1978). Health care practice is a subset of culture which is originated indigenously. Beliefs about the body and causation of illness, together with social norms and beliefs concerning when, why, and who to seek for medical help comprise one's "culture of medicine". Every society has a particular culture of medicine, which forms the culture's medical common sense, or logic. Therefore, it could be assumed that there are as many unique medicinal perspectives, as there are different cultures and subcultures. Approximately 80.0 percent of the world's population depends on traditional medicinal practices for their primary health care needs, as estimated by WHO.

The indigenous health care practices of North-east India are not well documented systematically; besides, it is the richest home for ethnomedicinal plant diversity in India. The traditional health care practice systems have survived since time immemorial among the people of north-east India. It is generally passed over generations orally and also through their personal experiences. Preservation of such traditional skills and knowledge perhaps required proper documentation and in-depth anthropological studies among different ethnic groups to explore the unexplored knowledge which is inherited culturally. Given the significance of traditional knowledge on the health care practice system, the present study was conducted to understand the indigenous health care practices of the Meiteis of Hilghat village, Manipur.

The land and people

The name of the village "Hilghat" is a Bengali word that means stones (*nungmacha* in Manipuri). The village is called The Hilghat village because of the abundance availability of small stones called "*nungmacha*" in Manipuri. The village was first settled by only eight surnames of the Meitei community such as Thiyam, Aheibam, Lairikyengbam, Paonam, Laishram, Wahengbam, Leihaorambam, and Thingbaijam as narrated by one of the eldest people of the village. The village is located 217 km west of Imphal, the state capital of Manipur. It is situated 4½ km away from the Jiribam sub-division, and it is the biggest village by area in the sub-district. The Hilghat village is surrounded on the eastern side by Khunou, on the western front by Kamaranga, on the northern side by Jiri River, and the southern side by Harinagar.

According to the household survey conducted in the present study, altogether there are 134 households in the village. The village has a total population of 660 and the number of males and females is found to be 336 (50.90%) and 324 (49.09%) respectively. The sex ratio of Hilghat village is 103 (assessment year April 2019).

Concept of illness

Traditional folk medicine is the substance of all the viewpoints on sickness and the healing methods applied against a disease that exists among the people. It is perceived that every illness has more than one potential cause and description. Poor health leads people to speculate about how they got sick, how it will affect them, and what they can do to make it better. People thus, form an explanatory model of the illness by integrating personal thoughts and circumstances with the popular illness ideologies of their culture. Explanatory models may vary from individual to individual. It may also differ from one culture to another. For example, in Hilghat village a person with high blood pressure is advised to eat one tamarind daily to control high blood pressure conditions. Moreover, it is also explained that the acidic nature of the tamarind will lower down the blood pressure and will not let it exceed further. And eventually, people with low blood pressure are not supposed to eat tamarind as it only helps in lowering blood pressure and it will cause serious damage to those who have low blood pressure. Also, the treatment of Jaundice, locally known as "*napu kabi*" which is a liver disorder, consists of the belief of pulling out the yellow liquid from all over the body through the one's sole or from the top of the head. They believe in taking out the undesirable yellow substance as a necessary means of curing the disease.

The villagers have the common basic idea of illness that people experience when there's an alteration of the natural process of the body, the mind, and the environment. They believe that causes of illness are attributed either by changes in the external physiological order or through evil eyes of the malevolent spirits, or witchcraft, and sorcery. Natural illness explains illness in impersonal systemic terms. Thus, the disease is thought to stem from natural forces or conditions such as cold, heat, and possibly by an imbalance in the primary body elements. The following form of illness or un-natural illnesses is related to the psychological substratum of the people, which has a footing on the magico-religious aspects. The un-natural or magico-religious concept of illness is attributed to the supernatural realm. It is caused by two major types of supernatural forces such as occult and spiritual causes. The occult cause is the result of evil spirits or human agents using sorcery while the spiritual cause is the result of penalties incurred for sins, breach of some taboos, or associated with a psychological social support system. The diagnosis of such illness is perceived and explicitly interpreted by an adept.

Indigenous health care practice

The Meiteis of Hilghat village practices traditional medicine and treatment of ailments with their knowledge of health and medicine inherited and learned by living in their culture. People of Hilghat village have rich indigenous health care practices used for treating various ailments. The traditional method of treating health problems seems to be less practiced to this day, but the tradition is still prevalent among the Meiteis of Hilghat village.

Their knowledge of using various herbs or plants for the treatment of disease has been handed down from generation to generation since time immemorial (Chopra *et al.,* 1956; Kirtikar and Basu 1975). A good number of medicinal plants are used which are either presented by the medicinal healers and also through people's perceptions of the plants that have medicinal values (Banerjee and Jalota 1988). However, the illness caused by supernatural power can only be treated by ablation, offerings, incantations, etc. while diseases caused by an imbalance in our body can be cured through medicines.

The ethnomedicinal specialists

There are two kinds of specialists in two different spheres of expertise. One is the "*Maibi*" or the traditional medicine practitioner or the sacred specialist who cures the health problems with the spiritual or supernatural power vested on her by her guardian force or god. And the other specialist is a medicine man. The Hilghat village has no traditional specialist that cures illness professionally but there is an old man in the village having ethnomedicinal knowledge who practices it when people consult him. The former specialist, "*Maibi*" is preferred mostly by the villagers and is frequently consulted for different issues related to health.

The medicinal healer

The Hilghat village has an elderly man of around 80 years of age named SM. Singh, who is well versed with the traditional knowledge of ethnomedicine. He does not practice it professionally but has equipped himself with the traditional way of healing using medicinal plants. He has learned about natural medicine from his forefathers and from many other old wise men of those times and has kept alive the tradition. He knows certain incantations used while treating a person with ailments since sometimes it requires incantations for better effectiveness of the traditional medicines.

The traditional medicinal specialist - *Maibi*

In the Hilghat village, a woman of middle age named PN Devi is the traditional medicinal practitioner locally called "*Maibi.*" She has been a specialist for the past six years. She cures ailments, both natural health problems and illness caused by supernatural entities among the people of Hilghat village and also the neighbouring village. She is reputed and respected by the villagers and holds an essential place in the lives of the people. She is also locally known as "*Ema*" by the people who believe and seek her treatment.

The *Maibi* is, what we refer to as the traditional medicinal specialist who can connect to the supernatural realm. Like every other *Maibi*, PN Devi too has a guardian force or God, which favoured her and gave her this extraordinary life. It happened some five or six years ago that she became ill and suffered for about six months. She did not know what her illness was? She had headaches and uneasiness of the body; she also had many unusual moments of crying out loud, screaming, laughing for no prior reason or so. She hoped for it to end, but things got out worse, and all her family members too felt sick for one reason or the other. It was almost like some misfortune had fallen upon them. She then consulted a *Maibi* for examining her health problem and all that's happening with her. It was then told that her illness was not due to any health issues but rather due to a supernatural reason. She was informed that she had been specially favoured and to be possessed (*Kaidongba)* by the guardian force or god called "*Nongpok Ninghtou Epa.*" She knew nothing of these events and so, sought help from the *Maibi* she consulted. She was then directed to get herself a "*Guru*" or a guide, which is referred to as "*Guru Langba*". If not, she won't be able to understand what to do and what not to be. The "*Guru*" has to be a professional "*Pena Khongba Maiba*" which means a *Maiba* who plays the traditional instrument called "*Pena*". The *guru* in her case is P Kanto from Jaipur, Cachar. He taught her all the norms of rituals and sacred practices, dos and don'ts, and obligations.

PN Devi has a sacred place in her courtyard, which she called "*Epa gi Khufam*" where she does the rituals and praying. She is refrained from household chores like cooking, washing dishes, sweeping, etc. She also does not consume meat and eggs; neither does she eat anything at anybody's home. She ought to maintain her sacred role and duty

Methods of treatment

Treatment of illness observed in Hilghat village may be discussed under two forms, namely the natural and supernatural ways of treatment. The natural treatments are performed by using herbal medicines and the supernatural or ritualistic way of treatment exists in the form of incantations and spells.

Treatment by medicinal plants

Some of the ailments and their cure using local medicinal plants:

(a) For Jaundice, a creeper is locally known as '*Uchishoombal*' approximately one foot in length is ground with seven leaves of

Black tulsi or Ram tulsi and the juice is to drink three times a day. After taking out the juice the residue is not to be thrown away, but rather, it is to be applied on the forehead at night. It is done so because it is believed to pull away all the yellowness from the body.

(b) For Joint Pain, the lady's finger seeds are to be taken and ground with *Kala namak,* and it is to be massage onto the joints like the knee. Also, the clay left after making pots is collected and made to dry in the sun and ground into a fine powder. It is then used to massage onto the joints.

(c) Alopecia areata or locally known as "*Samjabi*" is the loss of hair in patches. For its treatment, one fermented fish "*Ngari*" and one betel nut is roasted and ground into powder form. Then, the powder is mixed with mustard oil. And, it is applied to the affected area three times a day.

(d) When a person suffers from a common cold, betel leaf along with ginger and basil leaves are to be chewed together. Another treatment is by adding white turmeric powder and honey to water and drinking it.

(e) For diabetes, a tree named *Shaamba*, which is believed to have reliable healing power is used for treatment. The bark of the tree is boiled and the juice is drink daily. The other option is that the bark can be dried, ground into powder, and consume as it is.

(f) For pimples and boils of various kinds, the seeds of the Cinchona tree are ground into a powder and applied. Also, the leaves can be boiled in water for steaming to cure blisters.

A list of medicinal plants, a brief method of their use, and photographs commonly practiced by the traditional healers and *Ema Maibi* of Hilghat village are demonstrated in table 13.1.

Table 13.1: Names of some local medicinal plants with their medicinal values used in Hilghat Village

Sl. No	*The local name of the plant*	*Common English name*	*Botanical name*	*Medicinal uses*	*Photo*
1.	Angkot	Madar	*Calotropis gigantea*	Latex with little oil is applied to ringworms	

2.	Bhubati	Creat or King of bitter	*Andrographis paniculata*	Leaves used in chronic fever	
3.	Chanam	Garlic	*Allium sativum*	Garlic juice is used for skin troubles and ear drops	
4.	Kafoi	Pomegra nate	*Punica granatum*	Leaves are used in dysentery. A decoction of seeds with honey used for jaundice, colic, and diabetes	
5.	Khujila	Nux Vomica	*Strychnos nuxvomica*	Boiled extract of leaves applied to allergic swellings of skin. Bark extract used in cough and chest congestion	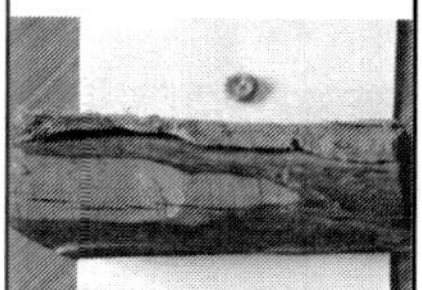
6.	Khongjai napi	Goat weed	*Ageratum conyzoides*	The plant used in local hair lotion "Chinghi". Leaves extract used in ringworm	
7.	Khujang	Balsam	*Impatiens balsamina*	Flowers are applied to burns. Leaves are applied on rotten nails	
8.	Langthrei	Ngai Camphor	*Blumea balsamifera*	Leaves useful in controlling blood pressure and gastric problems	
9.	Loiyumba lei/ loiyumei	Three leafed caper	*Crataeva nurvala*	Leaf juice used in stomach disorders and urinary complaints	
10.	Maroi napakpi	Hooker chives	*Allium hookerii*	Used for reducing high blood pressure	

11.	Neem	Margosa tree	*Azadirachta indica*	Leaves are antiviral, used in skin disease	
12.	Nongmang kha	–	*Phyloga-canthus thyrsiflorus*	Leaves used in cold and cough	
13.	Perook	Indian pennywort	*Centella asiatica*	Fresh plant juice is given with honey in an ulcer. Useful in digestive complaints	
14.	Poongtol	Guava	*Psidium guajava*	Young leaves given in dysentery	
15.	Ram tulsi	Shrubby basil	*Ocimum gratissimum*	Leaf juice mixed with honey is given in fever and cough	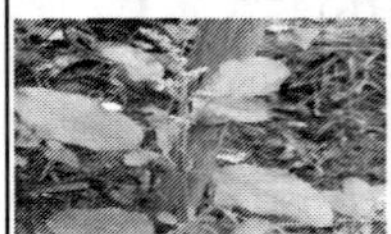
16.	Sanarei	Marigold	*Tagetes patula*	Crushed leaves are applied to fresh wounds for bleeding to stop	
17.	Shaamba	Midnight horror/ Indian trumpet flower	*Oroxylum indicum*	The boiled bark of the tree is taken for diabetes. Leaves used in muscular pain and general weakness	
18.	Tikhao yaikhu	Wild turmeric	*Curcuma aromatica*	Used in stomach disorders, cough, and womb problems	
19.	Tera paibi	Hill Gynura	*Gynura cusimbua*	Stem and leaf juice applied to a fresh wound to stop bleeding and quick healing	
20.	Uchisuzm bal	False daisy	*Eclipta prostata*	Used in jaundice or liver enlargement	

Treatment by oblation and incantations

The above treatment method by using medicinal plants is sometimes necessary to give "*wahei sonba*" which is a spell or incantation for better and faster results.

(a) For bleeding problems where clotting of blood is slow, crushed leaves of "*Tera Paibi*" (Goat weed) is applied on the wound and a "*wahei sonba*" which is read out as follows - "*Emu o - enou o - engang o - seloi ee - lamngai ee - leirang ee - tatna ee - ee taret asigi maming di durga kouwi - nacha taret makpu khambiyu*". Another incantation is the reverse names of the days of a week which is "*Thangja – erai – sagolsen – yumsakeisa – leibakkpokpa - ningthoukaba - nongmaijing*". It is to be repeated three times while applying the *Tera Paibi* paste on the wound to stop bleeding. It is believed that the reverse reading of the days of the week will also reverse the bleeding that it will clot immediately.

(b) The medicine man also treats some female health issues like white discharge by giving incanted water, without the use of medicinal plants. It is also sometimes provided with "*Jantra*" which is a talisman, to be worn over by a thread around the upper arm or waist. Headaches, eye problems, and nose bleeding are some of the ailments treated with spells. Such spells are not disclosed to people because it has too much information that needs to be shared.

(c) The traditional specialist *Maibi* treats the ailments of people by doing rituals, and by providing a specially incanted water "*Sida hidak*". Individuals with any illness about typical health disorders or of the unnatural illness can be treated. People come with their health complaints to her; she then performs rituals in her sacred place "*Epa gi Khufam*". She prays to the "*Nongpok Ningthou Epa*" to give her the cure. She is then directed by her "*Epa*" as follows "*Heiram taret leiram taret dhuk taret meira taret taret taret na khujing oiraga taibang migi anabado kokthok u chamthok u aduga Epa da sida hidak nirak u.Epa na sida hidak pinni*" which means that by offering seven kinds of fruits, seven kinds of flowers, seven incense sticks, and seven candles; she should wart off the ailments and purify their body, and then she should ask for the "*Sida hidak*" from "*Nongpok Ningthou Epa*". Here the number seven has been given a special or sacred value because the seven clans had formed the Meitei community. And, according to the directions from her

Epa, she incants the "*wahei sonba*" on water and gives it to the patients for treatment of various illnesses. For health issues relating to painful joints of the bone and muscle pain; incantation of *wahei sonba* is done on mustard oil or ghee and is given for massage. Unlike the medicine man, she has no pre-acquired knowledge about the cure for ailments.

Case study of patients

A man of 70 years, named KNS of Hilghat village, had suffered from boils along with fever for about three to four days. As there is no nearby hospital, it was a better choice to seek the traditional healing method. So his wife decided to consult *Ema Maibi* and narrated about the ailment. After that, the *Maibi* did the rituals by offering fruits, flowers, candles, and incense sticks to her god and asked for a cure. She then gave "*Sida Hidak*" which was incanted water based on the directions given by her god. She prescribed him to drink the "*Sida Hidak*" twice a day before meals for three to four days. Also, for the external boils, a paste made out of "*Lai perook*", marigold and '*langthrei*' were given to apply. And after the said days of medication, the patient was fully cured, and no such things happened again.

In another case, a woman named KA Devi had suffered from pain in the knee joints. She is around 60 years old, and it happened one year ago. She, too, did not consult a doctor because going to the hospital was a concern of time and money. She thought of bearing the pain with the false hope that the pain will go away soon on its own. But due to everyday household chores with frequent sitting and standing made it worse. So she consulted the *Ema Maibi* to treat her ailment. The necessary offerings and rituals, as stated in the previous case, were carried out. She was also given the "*Sida Hidak*" and had to drink it every day for two to three days. She also incanted on mustard oil and was prescribed to apply and massage it on the knee joints for a week. She followed the medication as instructed, and now the joint pain is cured.

A woman named Mrs. B, who is 61 years of age in the Hilghat village, suffered from Dysuria (locally known as *Esing yafam houba*), which is the difficulty or the painful discharge of urine. It so happened some three years ago. She stayed untreated for two days, but then the discomfort increased. She consulted the traditional healer S Manibabu. She was given a juice by boiling the bark of a tree called "*Shaamba*" which is proved to have much healing power. Soon her discomfort was cured after drinking the juice for three days.

Summary

The Meities of Hilghat village has rich living traditional knowledge of health care practices of its own learned from their forefathers. Such knowledge and skills were closely associated with their culture. Most of the people of the village believe in traditional health care practices besides accessing modern medicinal health care facilities. There are several treatment methods for different illnesses such as using varieties of medicinal herbs along with ritualistic performance, and incantations performed by the traditional specialist *Ema Maibi*. The present study indicates the importance of traditional knowledge of health care practices in healing ailments as well as provides satisfaction physically, mentally, and spiritually. Such pieces of knowledge and practitioners are occupying a significant role in the field of health care services in today's modern society.

The methods of treatment are eco-friendly, and it would also help in bringing sustainable livelihood. However, the age-old traditional health care services are fading down due to development in modern sciences and rapid changes in the lifestyle of the people as a result of modernization and globalization. Therefore, it is time to preserve the rich indigenous knowledge and skills of health-seeking behaviours associated with different cultures before it vanishes from our society. Simultaneously, the conservation of diverse biodiversity of medicinal plants also requires immediate attention. It could be achieved through undertaking in-depth anthropological studies on the indigenous health care system of different ethnic groups of India in general and north-east India in particular.

Acknowledgment

The authors would like to thank Saikhom Manibabu Singh (traditional healer) and Pounam Ngoubi Devi (Ema Maibi) and all the respondents for their bits of help and supports. The outcome of the present study would not have been possible without their support.

References

Banerjee, BG, and Jalota, R 1988, *Folk Illness, and Ethnomedicine*. New Delhi: Northern Book Centre.

Chopra, RN, Nayar, SL, and Chopra, IC 1956, *Glossary of Medicinal Plants of India*. New Delhi: CSIR.

Foster, GM, and Anderson, BG 1978, *Medical Anthropology*. Toronto: John Wiley and Sons.

Kirtikar, KR, and Basu, BD 1975, *Indian Medicinal Plants*. New Delhi: Periodical Book Express Co.

WHO (2001). http://www.who.int/topics/traditional_medicine/en/

Chapter - 14

Health and Nutritional Status among the Hmar Children of Churachandpur District, Manipur

J.W. Dkhar and Nancy Malsawmkim

Introduction

Health has been regarded as one of the most fundamental facets of human life. It represents not only the bio-social development of a person but also society as a whole. According to the World Health Organisation 'health is a state of complete physical, mental, and social well-being and not merely the absence of diseases or infirmity' (WHO,1971). Improvement in the health status of the population has been one of the most important objectives of a country's planning towards its overall development. The importance of health status studies can never be put aside; as such studies reveal certain important, fundamental facts that reflect the image of the health and nutritional status of a country or region in general, as well as the children of that region in particular. Although there have been improvements in some health indicators in the last two decades, including a reduction in child mortality rate and a drop in the fertility rate, improvements in nutritional status, have been less impressive (IIPS, 2007). The high prevalence of undernutrition among children is considered to be a major public health problem in rural India (Bose *et al.,* 2007). Despite the knowledge that discrimination in health care and nutrition may lead to severe growth faltering and increased mortality, people tend to discriminate. (World Bank, 1991).

Nutrition is the science of food and its relation to health. Nutritional status is the physical expression of the relationship between the individual dietary intake, the bioavailability of ingested nutrients, and his or her physical requirement (Brown,1984) influenced by the availability of food and purchasing power (Rao *et al.,* 2006). Food plays an important role in individual health as well as in disease. With the current transition from primitive to modern technology, there is an increase in lifestyle disorder, it is, therefore, important to promote healthy nutrition in all age groups. Reviews on the nutritional assessment conducted among the children and adolescent of the country have reported that the prevalence of overweight and obesity is observed among children and adolescents in urban India (Chhatwal *et al.,* 2004; Tharkar and Vijay 2009; Goyal *et al.,* 2010; Thakre *et al.,* 2011; Gupta *et al.,* 2013; Kumar *et al.,* 2014). Malnutrition, impairment of health either from a deficiency or excess or imbalance of nutrients, continues to be one of the major health problems, which is, of public health significance all over the world. It creates lasting effects on the growth, health, development, and physical fitness of a person. It is significantly associated with morbidity and mortality early in life and has long-term consequences, although it can occur anytime during the individual lifecycle (Malina *et al.,* 2004). Undernutrition continues to be the main cause of ill health and premature mortality among children and adolescents (Black *et al.,* 2000; Bose and Bisai 2008; Nandy *et al.,* 2005; Pelletier,1994). Developing countries like India continue to have a high prevalence of childhood undernutrition, mortality, and morbidities (Nandy *et al.,* 2005; Bamji *et al.,* 2008; Varadharajan *et al.,* 2013). Malnutrition, which is an impairment of health either from a deficiency or excess or imbalance of nutrients, continues to be one of the major health problems, which is of public health significance all over the world. It creates a long-lasting effect on the growth, development, and physical fitness of a person. It is significantly associated with morbidity and mortality early in life and has long-term consequences, although it can occur anytime during an individual life cycle (Malina *et al.,* 2004).

Therefore, the objective of the present study is an attempt to assess the health and nutritional status of the Hmar children age 3 to 9 years using nutritional anthropometry.

Material and methods

The present cross-sectional study was conducted among the Hmar children, a major tribal population of India, aged 3-9 years of Churachandpur district. The district occupies the south-west part of Manipur state. It stretches between 93 d° 15, E and 94°45, E Longitude and between 24°N and 24°30N Latitude. Churachandpur is the district headquarters of the largest district in Manipur. It covers an area of 4,750 kilometers. The term 'Hmar' is believed to have been originated from the term 'Hmarh' meaning 'tying of one's hair' in a knot on the nape of one's head. The Hmar's are physically strong and sturdy race, belonging to the Mongoloid stock, with a flat broad nose and short stature, and well-developed legs. The skin is usually yellow-brownish in colour. The face is broad and flat and the body is fairly well built with scanty body hairs. The men seldom have hair on their faces. (Allen *et al.,* 1979). The data collected from February 2019 to September 2019. The sample collection was conducted in the villages inhabited by the Hmar tribal population in the Churachandpur district of Manipur. General demographic and socio-economic data which includes relevant information on education, occupation house-type, source of drinking water, food habits, the energy source for cooking, family type, etc. were collected by following observation and interview method.

The anthropometric measurements of height and weight were recorded for children using standard procedures (Weiner and Lourie, 1981). The height of the children was recorded with the help of an anthropometric rod with the head of the subject being held in the Frankfurt Horizontal plane. Height was measured to the nearest 0.10cm. The weight of the children wearing minimum clothing and with bare feet was recorded using a portable weighing machine.

For assessment of nutritional status, three indices were used viz., weight for age, height for age, and body mass index for age. The body mass index was calculated as weight in kilogram divided by the square of height in meter. The assessment of nutritional status for the weight for age, height for age, and body mass index for the present study is expressed in terms of Z-scores and is calculated using LMS methods (CDC, 2000).

Results

From the present study, it is observed that the mean body weight and height are lightest and shortest at the lowest age group of 3 years (15.32 kg and 97.46cm) and tends to increase with the increase in age. Children falling under the age group of 9 years tends to be the heaviest and tallest (26.77 kg and 125.34cm) with regards to BMI, it tends to fluctuate in all the age group between 3 to 9 years. It has been observed that it is highest at the age of 9 years and lowest at the age of 6 years (Table 14.1 and Figures 14.1, 14.2, and 14.3).

Table: 14.1. Mean weight, height, and BMI concerning the age of the Hmar children

Age group (in years)	*N*	*Weight(kg)*		*Height(cm)*		*BMI(kg/m²)*	
		Mean	*SD*	*Mean*	*SD*	*Mean*	*SD*
3	38	15.32	2.12	97.46	5.37	16.14	2.01
4	37	16.67	2.17	101.06	5.58	15.76	2.13
5	40	17.04	2.61	105.79	6.36	15.19	1.44
6	36	18.78	2.10	112.95	6.01	14.73	1.39
7	33	20.93	3.05	117.66	4.58	15.22	1.60
8	31	23.25	4.06	122.30	5.30	15.11	1.53
9	34	26.77	7.08	125.34	8.27	16.89	3.31

N= Total No. of Populations; SD= Standard Deviation

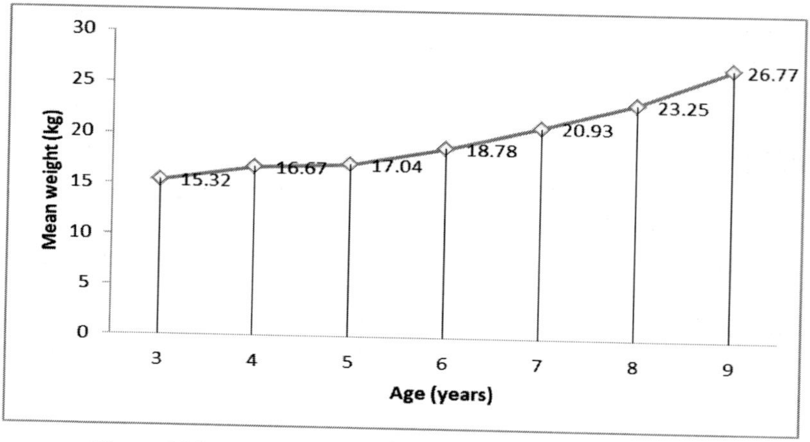

Figure 14.1: Distribution of the mean weight of the Hmar boys

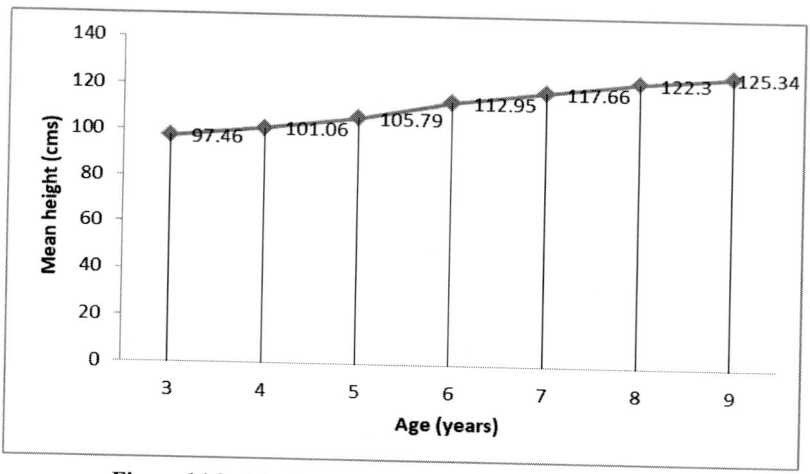

Figure 14.2: Distribution of mean height of the Hmar boys

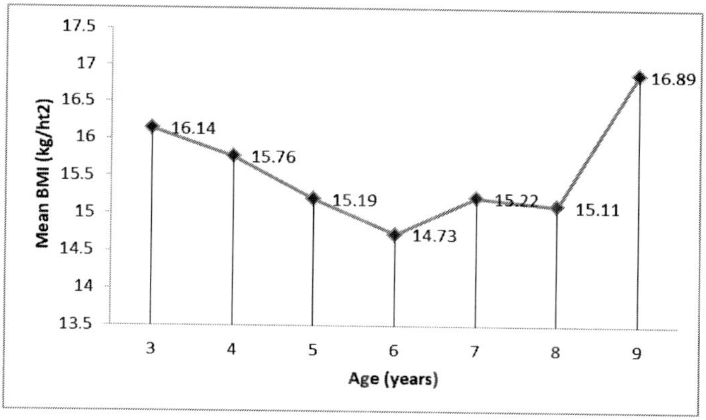

Figure 14.3: Distribution of the mean BMI of the Hmar boys

Height for age Z-score

Distribution of Height for age Z-scores of Hmar Boys in the overall age range from 3 to 9 years. It is seen from table 2 and figure 4, 77.52 percent of the boys are found to gain their height according to their age and 15.26 percent are found to be in moderate stunted condition while 7.22 percent of the boys are found to be in severe stunted condition.

Table 14.2: Distribution of height for age Z-scores of Hmar boys in the overall range from 3 to 9 years

Nutritional status	*Ranges*	*No. of individual*	*percentage*
Normal	≥2 Z-Score	193	77.52
Moderate	-2 to -3 Z-Score	38	15.26
Severe	≥-3 Z-Score	18	7.22
Total Z-Score		249	100

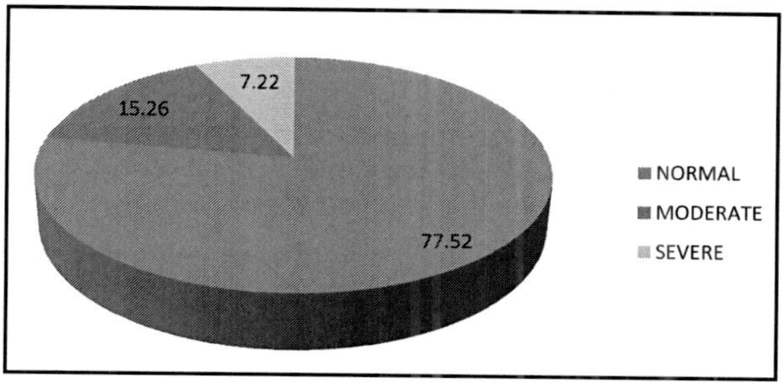

Figure 14.4: Percentage distribution of height for age Z-scored of Hmar boys in the overall range from 3 to 9 years

Weight for age Z-score

The nutritional status according to weight for age is shown in table 14.3 and figure 14.5. It is seen that 81.12 percent of boys in the present study are gaining weight according to their age. While 11.64 percent of boys are moderate underweight according to their age. 7.24 percent are found to be having a severe case which means they are severely underweight according to their age.

Table 14.3: Distribution of weight for age Z-scores of Hmar boys in the overall age range from 3 to 9 years

Nutritional Status	*Ranges*	*No. of individual*	*Percentage*
Normal	≥2 Z-Score	202	81.12
Moderate	-2 to -3 Z-score	29	11.64
Severe	≥-3 Z-score	18	7.24
Total Z-Score	-	249	100

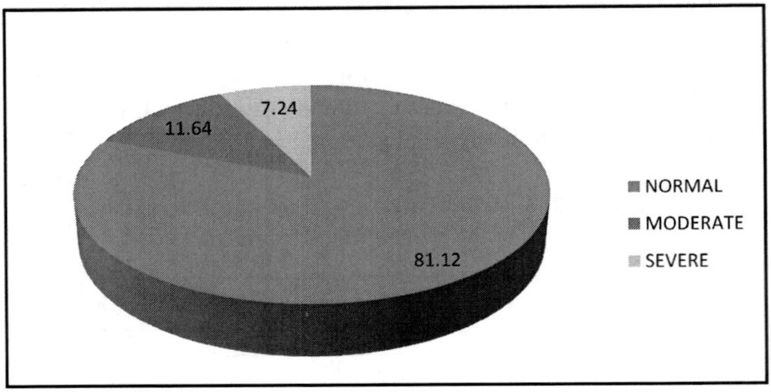

Figure 14.5: Percentage distribution of weight for age Z-score of Hmar boys in the overall range from 3 to 9 years

BMI for age Z-score

The distribution of children according to different levels of nutritional status concerning Body Mass Index (BMI), following the Z-score of -2 as a cut-off point is shown in table 14.4 and figure 14.6. From this table, it has been found that 86.34 percent are under normal condition, and 9.23 percent fall under the moderate condition, while 4.43 percent fall under the severe condition.

Table 14.4: Distribution of BMI for age Z-scores of Hmar boys in the overall range from 3 to 9 years

Nutritional status	*Ranges*	*No. of individual*	*Percentage*
Normal	≥2 Z-Score	215	86.34
Moderate	-2 to -3 Z-Score	23	9.23
severe	≥-3 Z-Score	11	4.43
Total Z-Score		249	100

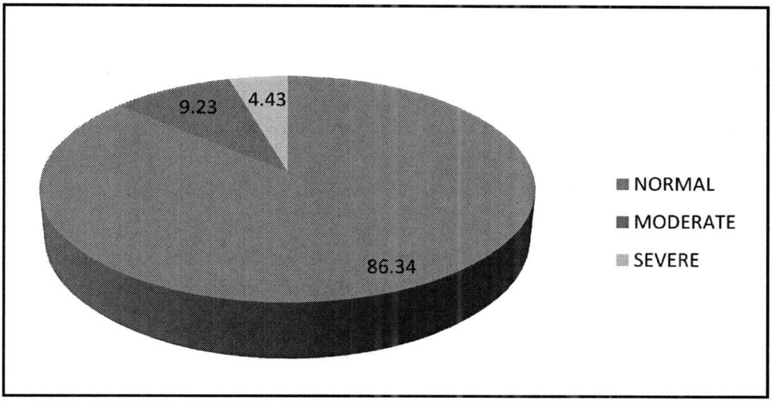

Figure 6: Percentage distribution of BMI for age Z-scores of Hmar boys in the overall age range from 3 to 9 years

Discussion

The weight and height of boys in the present study observed a gradual increase in trend with increasing age from 3 to 9 years, whereas, the BMI of the present study shows a fluctuate with different age groups. The BMI of the present study is found to be highest at the age of 9 years and lowest at the age of 6 years. Several cross-sectional studies conducted among different populations have also shown a similar trend, that both weight and height increase with the increase in age. Since the anthropometric examination is an appropriate means in any research to evaluate health and nutritional condition and the indices do not only reflect the socio-economic status of the family, health and social well-being of the population but also the competence of the health care system and the influence of the immediate surrounding can also predict the academic performance (Srivastava *et al.*, 2012). In the present study, three anthropometric indices have been taken, namely, height for age, weight for age, and BMI for age. The weight for age is considered for underweight and overweight while the height for age as an indicator of growth retardation and stunting growth and BMI for age is a good indicator of fatness and thinness or wasting due to Chronic Energy Deficiency. It is observed that in the present studied population, the prevalence of underweight, that is, weight for age is found to be 11.64 percent in moderate condition and 7.24 percent severe, it is found to be much lower than that reported from other parts of India, including North-east India (Medhi *et al.,* 2007; Rao *et al.,* 2006; Singh and Mondal, 2013). It has also been observed from the study population, 15.26 percent are moderately stunted and 7.22 percent are severely stunted. The cause of growth retardation is deeply rooted in poverty and lack of education (Maken and Varte, 2012) in their nutritional study of the Hmar tribe of North-east India. It is observed that children in low-income groups have a greater risk of stunting as compared to those in the high-income group. Rustein (2000) reported that the health status of young children and adolescents in developing countries is influenced by a wide variety of biological, behavioural, and socio-economic variables. Studies have also reported that in the last two decades, the dual burden of both underweight and overweight conditions co-existed together in many developing countries including India. Although underweight condition remains a major contributing factor (Jeemon *et al.,* 2009; Iyer *et al.,* 2011). The present studied population has also reported the prevalence of Energy Deficiency as measured by BMI is found to be 9.23 percent is under the moderate condition and 4.43 percent are severely affected. The possible reasons might be due to poverty that contributes to household food insecurity, poor maternal and childcare, unhealthy environments and poor access to health services, low dietary intake, and chronic infections (Ramzan *et al.,* 2008, WHO, 2011). According to Freedman and Perry (2000) and Kopelman (2000) that the lower values of BMI are associated with morbidity and also increase the expectation of health risks.

Conclusion

It may be concluded that although there is an existence of underweight and stunted growth among the Hmar children, the nutritional status of the present study is found to be better than many other populations from other parts of North-east India. Lack of awareness, education, poor health facilities, poverty has been contributed to the cause of undernutrition in the area. Therefore, proper nutrition, dietary practices, and health education are important to overcome this problem as the present population is undergoing an early phase of nutritional transition. It is also important to take preventive measures early in life, as the health problem early in life may have long-term consequences that may likely to persist into adulthood.

References

Allen, BC, Gait, EA, Allen, CGH & Howard, HF 1979, *Gazetteer of Bengal and North- East India*. New Delhi: Mittal Publications.

Bamji, MS, Murthy, PVVS, William, L & Vardhana, RA, MVV 2008, 'Maternal nutrition status and practices and perinatal neonatal mortality in rural Andhra Pradesh, India'. *Indian Journal of Medical Research,* vol.127, pp.44-51.

Black, RE, Morris.SS & Bryce, J 2000, 'Where and why are 10 million children dying every year? *Lancet,* vol. 361, pp. 2226-2234.

Bose, K & Bisai, S 2008, 'Nutritional status of rural adolescent school children in Paschim Medinipur, West Bengal'. *Indian Pediatrics,* vol.6, pp.515-516.

Bose, K, Biswas, S, Ganguly, S, Khatun, A, Mukhopadhyay, A & Bhadra, M 2007, 'Stunting underweight and wasting among Integrated Child Development Services(ICDS) scheme for children aged 3 to 5 years of Chapra, Nadia district, West Bengal, India'. *Maternal and Child Nutrition*, vol. 3, pp. 216-222.

Brown, NK 1984, 'Measurement of dietary intake'. *Population and Development Review,(Suppl),* vol. 10, pp. 69-91.

Center for Disease Control and Prevention 2000, National Health and Nutrition Examination.

Chhatwal, J, Manorama, V & Sandeep, KR 2004, 'Obesity among pre-adolescent and adolescents of a developing country(India)'. *Asia Pacific Journal of Clinical Nutrition,* vol. 13, no. 3, pp. 231-235.

Freedman, DS & Perry, G 2000, ' Body composition and health status among children and adolescents.' *Preventive Medicine*, vol. 31, no. 2, pp. 34-53.

Goyal, RK, Vitthaldas, S, Banshi, DS, Sanjiv RP, Navneet, NS, Mukesh, CG, Prashad, BR & Snehal, SP 2010, 'Prevalence of overweight and obesity in Indian adolescent school going children: Its relationship with socio-economic status and associated lifestyle factors'. *Journal of Associations Physicians of India,* vol.58, pp.151-158.

Gupta, R, Raisana, SK, Acharya, A & Bachani, D 2013, 'Socio-demographic co-rrelates of overweight and obesity among adolescents of an urban area of Delhi, India'. *Indian Journal Of Community Health,* vol. 25, no. 3, pp. 238-243.

IIPS 2007, *International Institute for Population Sciences and Macro International, National Family Health Survey(NFHS-3),2005-2006:India*: Volume I. Mumbai International Institute for Population Sciences.

Iyer, UM, Bhoite, RM & Roy, S 2011, 'An exploratory study on the nutritional status and determinants of malnutrition of urban and rural adolescent children(12-16) years of Vadodara city.' *International Journal of Applied Biology and Pharmaceutical Technology*, vol. 2, no. 1, pp. 102-107.

Jeemon, P, Prabhakaran, D, Mohan, V, Thankappan, KR, Joshi, PP, Ahmed, F, Chaturvedi, V & Reddy, KS 2009, 'Double burden of underweight and overweight among children (10-19 years of age) of employees working in Indian industrial units.' *The National Medical Journal Of India,* vol. 22, no. 4, pp. 172-176.

Kopelman, PG 2000, 'Obesity as a modern problem.' *Nature,* vol. 404, pp. 635-643.

Kumar J, Sangeeta, B, Joag, GG & Basveshwar, P 2014, 'Correlates of body mass index and waist circumference with blood pressure in school-age children'. *International Journal of Recent Trends in Science and Technology,* vol. 11, no. 1, pp. 109-112.

Maken, T & Varte, LR 2012, 'Nutritional status of children as indicated by Z-scores of the Hmars: A tribe of North-east India.' *Antrocom Online Journal of Anthropology*, 8, pp. 213-237.

Malina, RM, Bouchard, C & Bar-Or, O 2004, *Obesity in childhood and adolescence. In: Growth, maturation, and physical activity.* Second edition. USA: Human Kinetics, pp. 527-551.

Medhi, GK, Mahanta, NC & Mahanta, J 2007, 'Nutritional status of adolescents among tea garden workers.' *Indian Journal of Pediatrics,* vol. 74, pp. 343-347.

Nandy, S, Irving, M & Gordon, D 2005, 'Poverty, child undernutrition, and morbidity: New evidence from India'. *Bulletin of the World Health Organisation,* vol. 83, pp. 210-216.

Pelletier, DL 1994, 'The relationship between child anthropology and mortality in developing countries: Implications for policy, progress, and future research'. *Journal of Nutrition,* vol. 124, pp. 2047-2081.

Ramzan, M, Ali, I & Khan, AS 2008, 'Body mass status of school children of Dera Ismail Khan, Pakistan.' *Journal of Ayub Medical College Abbottabad,* vol. 20, no. 4. pp. 119-121.

Rao, KM, Kumar, RH, Venkaiah, K & Brahman, GNV 2006, 'Nutritional status of Saharia-A primitive tribe of Rajasthan'. *Journal of Human Ecology,* vol. 19, no. 2, pp. 117-123.

Rustein, SO 2000, ' Factors associated with infant and child mortality in developing countries in the 1990s'. *Bulletin of the World Health Organisation*, vol. 78, pp. 1256-1270.

Singh, J & Mondal, N 2013, ' Assessment of nutritional status: A case of tribal children in Assam, North-east India.' *Journal of Nepal Paediatric Society*, vol. 33, no. 1, pp. 1-7.

Srivastava, A, Mahmood, SE, Srivastava, PM, Shrotriya, VP & Kumar, B 2012, 'Nutritional status of school-age children- A scenario of urban slums in India'. *Archives of Public Health,* vol. 70, no. 8.

Thakre, SB, Mohnae, SP, Ughade, SM, Morey, SS & Humne, AY 2011, 'Correlates of overweight and obesity among urban school going children of Nagpur city'. *Journal of Clinical and Diagnostic Research,* Vol. 5, no. 8, pp. 1593-1597.

Tharkar, S & Vijayan, V 2009, 'Impact of socio-economic status on prevalence of overweight and obesity among children and adolescents in urban India'. *The Open Obesity Journal*, vol. 1, pp. 9-13.

Varadharajan, KS, Thomas, T & Kirpa, AV 2013, 'Poverty and state of nutrition in India'. *Asia Pacific Journal of Clinical Nutrition,* vol. 22, pp. 326-339.

Weiner, JS & Lourie, JA 1981, *Practical Human Biology.* Academic Press. London.

World Bank 1991, *Gender and poverty in India: A world bank country study.* Washington DC: The World Bank.

World Health Organization 2011, C*hildhood nutrition and progress in implementing the international code of marketing of breast milk substitutes.* WHO, Geneva, EB 109/11.

Chapter - 15

HIV and Tribal Youths: A Study in Jaintia Hills, Meghalaya

Roumi Deb

Introduction

Human Immunodeficiency Virus, HIV, which results in Acquired Immuno-Deficiency Syndrome (AIDS), has taken the lives of millions of people in the world and yet approximately 36.7 million people were living with HIV at the end of the year 2015 (WHO, 2017). There are various sources of HIV infection, including contaminated blood and blood products (such as shared medical injections, blood transfusions, and injection drug usage [IDU]), an infected mother transmitting the virus to her baby (before, during, or after birth and through breast milk), and through unprotected sexual relationship (Cohen *et al.,* 2008). Due to the risk behaviors, some high-risk groups have been identified, such as female sex workers, men who have sex with men, injecting drug users, young women, and children. Various factors like lack of knowledge about HIV/AIDS, inaccessibility to healthcare services and commodities, lack of education and life skills, and early marriage have increased the vulnerability towards HIV/AIDS (Kulkarni *et al.,* 2013).

HIV prevalence in India has been estimated to be 0.3 percent (equating to 2.1 million people living with HIV) and is the third-largest contributor to the global HIV epidemic (UNAIDS, 2016). The HIV epidemic in India is concentrated in nature, affecting mostly higher-risk subgroups (Phanuphak *et al.,* 2015). Besides, there is a regional variation in the prevalence of HIV, with the highest prevalence being reported from the North-eastern states of India (Sarkar *et al.,* 1993; Medhi *et al.,* 2012 and 2013; ISHA, 2014; Lucas *et al.,* 2015). Various epidemiological studies have been carried out in many parts of India, however, no data is available to highlight the knowledge and attitude of the tribal population of Jaintia Hills, Meghalaya. Therefore, the present study was carried out focusing on prevalence, knowledge, attitude, practice, and behavior on HIV/AIDS infection among the youths of Jaintia Hills, Meghalaya.

Materials and methods

This cross-sectional study was carried out amongst 1200 tribal youths belonging to the age group 17-24 years in Jaintia Hills, Meghalaya. Meghalaya is a North-eastern State in India constituting eight districts. The present study was conducted in the East and West Jaintia Hills districts of the state covering twenty-five villages from five blocks (Thadlaskein, Laskein, Amlarem, Saipung, and Khleriet). Random sampling was used for the recruitment of participants. The study population consisted of a total of 1200 tribal youths (males-520 and females-680; aged 17-24 years). The sample size of 1067 participants was calculated using an online sample size calculator (http://www.surveysystem.com/sscalc.htm) giving prevalence estimates with a 95.0 percent confidence level and within 3.0 percent confidence interval based on the total population of Jaintia Hills district i.e. 3,92,852 persons as per Census-2011. Before the data collection written permission was sought from Deputy Commissioner of East and West Jaintia Hills, District Medical and Health Officer (DMHO), and Medical officers (MOs) of all the health centers. Written consent was obtained from each patient before the conduct of the interview and information collected was kept confidential by the investigator. Patients who refused to give consent and who were unable to understand the questions were excluded from the study. A structured schedule was used to collect data on demographic data about their socio-economic status, educational and occupational status, marital history, living status, knowledge, attitude, perception about HIV/AIDS, and risk behaviors. In-depth Interviews and Focused Group Discussion (FGD) were used to collect qualitative data.

Results

The study found that the majority of youths belonged to the Pnar tribe (93.7%) and the remaining 6.3 percent belonged to the Baite tribe of East and West Jaintia Hills, Meghalaya. The subjects were mostly Christians (52.3%), Niamtres (26.4%), and Hindus (21.3%). About 91.0 percent of youths were pursuing their education. More than half (51.1%) of the youths were from rural areas and the rest were from urban areas (48.9%). The majority of them belonged to nuclear families (78.0%) and only 22.0 percent lived in joint families. Fifty percent of youths belonged to lower socio-economic status earning Rs 5000- Rs 10000 per month. The serological screening revealed the prevalence of HIV to be 5.4 percent (65 out of 1200 youths), with more females (57) than males (8) affected with the virus. The study revealed an association of HIV susceptibility with female gender, illiteracy, rural locality, and nuclear family type, showing statistical significance at a level of 0.05, after adjusting for confounding factors. Findings from Focus Group Discussion (FGD) revealed that poverty was the most important factor that had an impact on HIV/AIDS infection among the youths.

In the present study, it was observed that many of the young women belonging to the lower socio-economic status had multiple sexual partners as this was one of the means of earning their livelihood. Some of the girls were even forced to have multiple sexual partners in exchange for money by stepfather/stepmother. Some were found to be sexually abused (more than one time) and then they started having multiple sex partners. The majority of the Injecting Drug Users (IDUs) belonged to lower socio-economic status and they used to exchange their clothes, watches, or other goods to buy the drugs. Further estimation of knowledge about HIV/AIDS revealed that nearly two-thirds of the young general population heard of HIV or AIDS. The average knowledge score of 4.9 out of 10 found in the study suggested that the youths had nearly 50.0 percent knowledge on HIV (such as the physical appearance of HIV infected individuals, the consequences of infection, and its possible cure). Youths demonstrated some knowledge on the correct modes of transmission of HIV/AIDS like unprotected physical contact and homosexuality (74.0% - 76.0%), however, other main sources of infections were very less known like sharing sharp needles/injections or untested blood transfusion (2.0% - 12.0%). The average score on knowledge on HIV transmission mode was 55.0 percent (6.1 out of 11). Around 60.0 – 77.0 percent of studied youth had good knowledge about preventive measure (such as avoiding sharing razors, blades, or any sharp objects, avoid sex with homosexual, avoid sex with persons who have many sexual partners, avoid sex with multiple sexual partners, use a condom, abstain from sex) or perception on HIV/AIDS that can prevent them from the infection. The commonly mentioned preventive measures for HIV/AIDS infection reported in the FGDs were being faithful to partners, blood testing for HIV, using sterilized sharp objects. The findings of the FGDs further showed that school-going youths (15-18 years) were more knowledgeable regarding the transmission and preventive services of HIV/AIDS as compared to the local unmarried community youths

(17-23 years). This was because the unmarried youths had lower education as compared to the school going youths.

Youths were found to have many misconceptions such as getting HIV from public toilets or sharing food with an HIV-infected person. In the FGDs, some of the participants also reported that the sexual relationship through which HIV/AIDS infection is transmitted is something unavoidable and hence, they had nothing to do about it. However, attitude towards the use of condoms in the study was reported to be that 23.0 percent of the youths found it embarrassing to buy condoms and 77.0 percent believed that it reduces pleasure. It was also observed that women are scared that if they ask their partner to use condoms, then the man would be offended. There was a common myth among the youth in the study area as they believed that HIV/AIDS is a punishment from God. Respondents had a negative approach to HIV infected individuals and thought that they should be expelled from their respective communities.

The assessment of understanding related to HIV/AIDS among the 65 infected youths revealed that all the HIV infected youths knew the HIV or AIDS abbreviation and its consequences; however, the Anti Retro Viral Therapy (ART) was not known to 58.5 percent infected youths and alarmingly 95.0 percent were not receiving the therapy. The majority of infected youths (87.0 – 96.0%) knew about the sexual modes of transmission (unprotected sex, homosexual and anal intercourse); however, nearly 80.0 percent were aware that untested blood transfusion, infected needles/ injections/sharp objects could also spread HIV. Most of the HIV prevention methods (to be faithful to the partner, avoid unprotected sex, using condoms, etc.) were well known to the majority of infected youths. Moreover, there were no Anti Retro Viral Therapy (ART) centers reported in Jaintia Hills, so the patients visited the hospitals in the capital of the state (Shillong). It was found that the poor had less access to health care and, therefore, most of the HIV cases remain unregistered and untreated at the ART center resulting in further spread of infection.

Focus group discussions were conducted among the high-risk groups like injecting drug users and people who have multiple sexual partners. According to the opinion of the injecting drug users having sexual relations with multiple partners was the main reason for HIV/AIDS infection. In contrast to this, those who had multiple sexual partners thought that having more partners was not an issue but unsafe sexual practices were a major risk factor for HIV/AIDS infection. It was reported by the subjects that youths in the study region usually started taking drugs at the age of

15-16 years. The main reasons for taking drugs were depression, peer pressure, and being forced by partners. The study revealed that there is discrimination for the infected youths within their own families, workplaces, and health service delivery points.

The presence of HIV related risk behaviors like inconsistent use of a condom (66%), multiple sexual partners (28.3%), alcohol consumption before intercourse (22.3%), and domestic violence (5.0 – 10.0%) was observed. Findings from the FGDs also reported similar observations concerning risk behaviors. Unprotected sex, multiple sex partners, the increased establishment of temporary sexual relationships, consumption of alcohol, and drug use were the common risk behaviors among the youths in the region. FGDs also revealed a lack of knowledge regarding the proper use of condoms. Married women had no decision on safe sex even if they were aware that the husband practices extramarital relationships and could infect them also with HIV/AIDS. The perceived HIV/AIDS-related problems included the malicious spread of HIV/AIDS by some infected people and deliberate discrimination of people and families affected by HIV/AIDS. Findings also depicted that some people who know or suspect that they are infected want to take revenge by maliciously spreading HIV/AIDS. The results revealed that infidelity was reported by 28.0 percent of sexually active youths and they were aware of the infidelity of their regular sexual partner. The prevalence of domestic violence was reported by around 10.0 percent of the youths that included all three forms- sexual (4.9%), physical (9.2%), and verbal abuse (10.3%), posing a risk to the study population for HIV infection. Despite the presence of HIV risk behaviors, the personal perception of HIV risk was found to be only among 8% of the subjects. Although, only one percent (1.0%) of them thought that their partner's sexual behavior put them at risk of HIV infection.

Despite so many initiatives taken by central and state governments, only one percent of the youths were found to be aware of any programs aimed to prevent the spread of HIV or AIDS. The youths were further interviewed to estimate their awareness regarding five different schemes that are available/ operational in the area of study according to the records of the government of Meghalaya. Nearly 82 youths (6.8%) had some idea about one or the other schemes. Out of them, around 6% knew about the School AIDS Education scheme. Nearly 4 percent of youths were aware of Targeted Intervention Scheme; however, all of them were not sure of its existence in the study area. Only 1.0 percent of youths were aware of the National AIDS helpline and tele-counseling scheme. No one ever heard of the Community Care and Support scheme or Relief scheme for HIV/AIDS.

In the present study, various approaches were undertaken to create awareness for reducing the prevalence in the community. These included counseling at an individual level and for the high-risk groups (multiple sexual partners and intravenous drug users). The individuals were encouraged to visit health centers for HIV testing and counseling centers; reduce the number of multiple partners, maintain sexual hygiene, use a condom; abstain from sharing needles, syringes, and sharp object; clarify myths and misconceptions regarding spreading HIV/AIDS; abstain from drinking and smoking; provide clear information on free ART, etc. At the community level, focus group discussions were carried out to counsel and create awareness among the study population; staff and children at shelter homes were made aware about HIV/AIDS infection, its prevalence, transmission, and prevention methods; Self Help Groups (SHG) were created to increase the awareness in the local population.

Conclusion

The study has revealed a high rate of HIV infection among the tribes of Jaintia Hills, Meghalaya, with more females affected than males. The incidence is associated with a rural locality, nuclear families, and marital status. Moreover, overall knowledge and perception of the youths and children were not satisfactory for many of the variables like consequences of HIV/AIDS, modes of transmission and prevention of the spread of AIDS-causing virus, and HIV-related stigma. The situation further gets intensified with minimal health facilities related to HIV-testing, counseling, and treatment in the district. It is therefore suggested that HIV awareness initiatives to be conducted at the regional level, educate the community through social media, multi-media, street-plays, workshops, religious gatherings, etc, Improvements in health facilities like HIV-testing, proper counseling maintaining confidentiality, and development of ART center in the district would enhance the health-seeking behavior of the tribal population.

Acknowledgement

Gratefully acknowledge the financial support received from ICMR. Thanks to the District Authorities and youths of the study area and for their co-operation and support.

References

Cohen, MS, Hellmann, N, Levy, JA, De Cock, K & Lange, J 2008, 'The spread, treatment, and prevention of HIV-1: evolution of a global pandemic'. *The Journal of Clinical Investigation*, vol. 118, no. 4, pp. 1244-1254.

ISHA, 2014, HIV-AIDS *Situation in North-Eastern States. Indian* Society for Health Administrators. (www.medind.nic.in/haa/t04/i1/haat04i1p147.pdf).

Kulkarni, P & Vishwanath, P 2013, 'Adolescents and HIV: This is time to prioritize'. *International Journal of Health Allied Science*, vol. 2, pp. 223-224.

Lucas, GM, Solomon, SS, Srikrishnan, AK, Agrawal, A, Iqbal, S, Leyendecker, O, McFall, AM, Kumar, MS, Ogburn, E, Celentano, DD & Solomon, S 2015, 'High HIV burden among people who inject drugs in 15 Indian cities'. *AIDS,* (London, England), vol. 29, no. 5, pp. 619.

Medhi, GK, Mahanta, J, Kermode, M, Paranjape, RS, Adhikary, R, Phukan, SK & Ngully, P 2012, 'Factors associated with history of drug use among female sex workers (FSW) in a high HIV prevalence state of India'. *BMC Public Health*, vol. 12, no. 1, pp. 273.

Medhi, GK, Mahanta, J, Hazarika, I, Armstrong, G, Adhikary, R, Mainkar, M & Paranjape, RS 2013, 'Syphilis infection among female sex workers in Nagaland, North-east India: analyzing their vulnerability to the infection'. *International Journal of STD and AIDS*, vol. 24, no. 3, pp. 193-199.

Phanuphak, N, Lo, YR, Shao, Y, Solomon, SS, O'Connell, RJ, Tovanabutra, S, Chang, D, Kim, JH & Excler, JL 2015, 'HIV epidemic in Asia: Implications for HIV Vaccine and other prevention trials'. *AIDS Research and Human Retroviruses*, vol. 31, no. 11, pp. 1060–1076.

Sarkar, S, Das, N, Panda, S, Naik, TN, Sarkar, K, Singh, BC, Ralte, JM, Aier, SM & Tripathy, SP 1993, 'Rapid spread of HIV among injecting drug users in North-eastern states of India'. *Bull Narc*, vol. 45, no. 1, pp. 91-105.

UNAIDS, 2016, *Report on the Global AIDS update*. Geneva: Joint United Nations Programme on HIV/AIDS; 2016.

World Health Organization, 2017, *Mother-to-Child Transmission of HIV* [accessed 17 April 2017].

Chapter - 16

Prevalence of Dental Caries among the Mizos of Aizawl Town, Mizoram

DHRUBA KUMAR LIMBU AND C. LALRAMMUANA

Introduction

Dental anthropology is a specialized field of physical anthropology under the broad discipline of anthropology which is concerned with the study of human teeth – one of the human anatomical systems. One of the main aims of dental anthropology has been a study of the process of wear and other changes that occur once they are in place.

Dental caries is the destruction of enamel, dentine, and cement resulting from acid production by bacteria in the dental plaque, ultimately leading to the formation of a cavity in the crown or root surface (Hillson, 1996). It is a pathological condition of the teeth resulting from the decalcification of the dentine or enamel and the disintegration of the remaining organic material often leading to the loss of the teeth. The International Dental Federation Commission on Classification and Statistic for Oral Conditions in its recommendation to the World Health Organization (1975) defined dental caries, "as a localized, pathological process of bacterial origin that results in the demineralization of the hard tooth structures and progression to cavitations. Caries, being a disease process, starts with a microscopic lesion and eventually progresses into a microscopic cavity". Dental caries occurs universally, though the severity and resulting problems vary from country to country, even from region to region within the same country.

Dental caries is said to be a disease of civilization (James, 1979). According to the studies conducted by Barrett and Williamson (1972), the incidence of caries increases with the advancement of time who studied the Yuendumu population twice at the interval of 18 years. The disease was studied to see the differences in prevalence by sex (Nelson, 1937-38 and Goldstein, 1948), age (Barnard, 1956), and race (Heitetz *et al.,* 1976).

In India, the existing literature shows that the works on dental caries are scanty. The pioneering works were done by Shourie (1946), Pal (1964), Rami Reddy and Naidu (1980), and Rami Reddy *et al.,* (1982b).

The present chapter on the prevalence of dental caries in the permanent teeth among Mizo is aimed at providing additional literature to the existing lot on the subject for India in general and North-east India in particular.

Materials and methods

The data for the present study was collected from Aizawl town of Mizoram. There are a total of 69 village councils in Aizawl town, out of which 7 were selected by applying a simple random sampling method. The subjects included for the present study were 5151(M = 2409; F = 2742) individuals of all the age groups belonging to the Mizo tribe of Aizawl town, Mizoram.

The term 'Mizo' is a collective name for the indigenous people of Mizoram, possessing one language, same origin, and a common way of life. The Mizos belong to the Mongoloid race and Tibeto-Burman linguistic group. Etymologically, the word 'Mizo' means 'Hillman' which is derived from the two Mizo words, viz., *Mi,* and *Zo*, meaning man and hill respectively. A Mizo family is constituted through the institution of marriage. The system that existed regarding the exercise of leadership in the family was patriarchal. In all affairs relating to family life such as; celebration, issues of inheritance, etc., the husband or the father exercises power. Regarding recognition of the descent of the married couple, the patrilineal system was in vogue, i.e., an heir was recognized through the male line (Nag, 1993).

The staple food of the Mizo is rice and millet. They are fond of meat and eat almost every kind of animal. They take different kinds of vegetables, leafy vegetables, and varieties of edible roots and herbs available in the *jhum* and forest. Locally made rice beer is their favourite drink. Cigarettes and pipes are very common in the village. Chewing of betel nut and betel leaf and tobacco is also common among the Mizos. However, due to the influence of Christianity and education, many of them stopped consumption of rice beer and smoking.

Since dental caries has to be assessed for the entire dentition, all the teeth of the subjects were examined. Following WHO's (1977) recommendations, the teeth in either type of dentition were examined and diagnosed sound whether they are un-affected by caries, and decayed, filled, and missing owing to caries. The subjects were examined at their houses in broad daylight with the help of a dental mirror, dental probe, spatula, and torch. The method of direct visual observation was followed during the investigation. The data collected were analyzed by age, sex, the position of teeth on the jaws, and bio-social traits.

Data on household income were collected directly from the heads of the households. The per capita monthly income of the households was classified as follows:

Below 50th percentile (below Rs. 5000) = Low-income group

50th to 75th percentile (Rs. 5000- Rs 10000) = Middle-income group

Above 75th percentile (above Rs. 10000) = High-income group

Results

The influence of certain biological and socio-economic factors such as income, education, and eating habits such as betel nut and betel leaf (*Kuhva*) chewing, tobacco chewing, smoking cigarette, and pipes on dental caries among the study population have been discussed in the following Tables. In respect of the position and age-group, wise distribution of caries among the males and females are also dealt with dental caries among the Mizos of Aizawl town, Mizoram.

Table 16.1: Caries affected individuals by age and sex

Age (year)	*Mild*		*Moderate*		*Severe*	
	Male	*Female*	*Male*	*Female*	*Male*	*Female*
Below 1	0	0	0	0	0	0
1.	0	0	0	0	0	0
2.	6(9.68)	7(11.11)	3(4.84)	2(3.17)	0	0
3.	7(12.96)	4(6.78)	4(7.07)	1(1.69)	0	1(1.69)
4.	11(17.19)	12(21.05)	3(4.69)	2(3.51)	2(3.13)	1(1.75)
5.	9(19.57)	10(25.64)	4(8.70)	2(5.13)	3(6.52)	1(2.56)
6.	7(18.92)	6(24.00)	4(10.81)	3(12.00)	3(8.11)	2(8.00)
7.	7(17.95)	5(10.64)	7(17.95)	3(6.38)	3(7.69)	3(6.38)
8.	6(11.11)	4(10.00)	3(5.56)	3(7.50)	2(3.70)	2(5.00)
9.	8(18.60)	6(13.33)	4(9.30)	4(8.89)	2(4.65)	2(4.44)
10.	6(18.18)	7(19.44)	3(9.09)	4(11.11)	4(12.12)	3(8.33)
11.	9(18.37)	8(17.78)	3(6.12)	2(4.44)	1(2.04)	1(2.22)
12.	6(14.29)	6(11.76)	2(4.76)	3(5.88)	2(4.76)	1(1.96)

(Cond...)

13.	7(20.00)	6(15.38)	4(11.43)	4(10.26)	3(8.57)	3(7.69)
14.	6(13.33)	7(13.21)	4(8.89)	3(5.66)	1(2.22)	1(1.89)
15.	5(11.36)	4(10.53)	3(6.82)	3(7.89)	2(4.55)	2(5.26)
16.	8(13.56)	6(8.96)	4(6.79)	5(7.46)	3(5.08)	2(2.98)
17.	7(14.89)	8(14.29)	6(12.77)	7(12.50)	5(10.64)	4(7.14)
18.	9(15.79)	10(14.93)	6(10.53)	5(7.46)	4(7.02)	4(5.97)
19.	11(16.67)	8(14.55)	5(7.58)	5(9.09)	4(6.06)	3(5.45)
20.	13(18.84)	11(16.92)	6(6.70)	7(10.77)	4(5.80)	3(4.62)
21.	11(17.74)	14(18.92)	6(9.68)	9(12.16)	5(8.06)	6(8.11)
22.	8(19.05)	16(21.33)	5(11.90)	9(12.00)	3(7.14)	7(9.33)
23.	12(19.05)	14(20.29)	7(11.11)	6(8.70)	5(7.94)	5(7.25)
24.	8(17.78)	13(19.12)	4(8.89)	5(7.35)	3(6.67)	3(4.41)
25.	9(20.00)	13(18.57)	6(13.33)	6(8.57)	3(6.67)	2(2.86)
26-30.	47(21.66)	54(19.22)	39(17.97)	47(16.73)	21(9.68)	31(11.03)
31-35.	34(17.26)	29(14.15)	26(13.20)	23(11.22)	19(9.64)	22(10.73)
36-40.	27(19.01)	31(17.32)	28(19.72)	33(18.44)	15(10.56)	17(9.50)
41.45.	19(16.24)	21(13.46)	14(11.97)	17(10.90)	14(11.97)	18(11.54)
46-50.	12(10.26)	14(10.47)	13(11.11)	16(11.94)	11(9.40)	15(11.19)
51-55.	8(10.13)	11(9.82)	6(7.59)	12(10.71)	6(7.59)	10(8.93)
56-60.	5(5.10)	5(5.43)	7(7.14)	6(6.52)	9(9.18)	8(8.70)
61-65.	3(5.88)	5(7.35)	4(7.84)	7(10.29)	7(13.73)	9(13.24)
66-70.	3(10.00)	3(9.38)	2(6.67)	3(9.38)	4(13.33)	3(9.38)
71+.	8(13.11)	8(11.59)	11(18.03)	15(21.74)	13(21.31)	17(24.64)

Figures in parentheses represent the percentage.

Table 16.1 shows the numbers and percentages of caries affected persons by age and sex. Caries affected teeth were first recorded at 2 years old children of both sexes. The severe form of caries is first observed at 4 years and 3 years in males and females respectively.

From table 16.2, it is observed that the percentage of caries affected individuals are highest in the age group, 26-30 years followed by 31-35 years, and then 36-40 years in both the sexes. Caries affected persons increase with the increase of age up to 31-35 years, thereafter, gradually decrease with the increase of age. In respect of moderate and severe forms of caries, the frequencies of males are higher than their female counterparts.

The prevalence of caries in deciduous and mixed dentition (deciduous and permanent teeth) by age and sex are presented in table 16.2. Mild and moderate forms of caries are recorded earliest at the age group of 13-18 months in both sexes. The highest percentage of a mild form of caries is observed at 49-54 months age in both the sexes, while for the moderate

Table 16.2: Prevalence of caries in mix dentition (deciduous and permanent teeth) by age and sex

Age (month and year)	*Mild*		*Moderate*		*Severe*	
	Male	*Female*	*Male*	*Female*	*Male*	*Female*
≤ 6 months	0	0	0	0	0	0
7-12 months	0	0	0	0	0	0
13-18 months	2(7.14)	3(8.82)	1(3.57)	1(2.94)	0	0
19-24 months	4(11.43)	4(12.12)	2(5.71)	1(3.03)	0	0
25-30 months	3(11.11)	3(10.00)	2(7.40)	2(6.67)	0	0
31-36 months	4(11.76)	3(11.11)	2(5.88)	2(7.41)	0	0
37-42 months	3(15.00)	4(12.50)	1(5.00)	1(3.13)	0	0
43-48 months	5(15.63)	3(10.00)	2(6.25)	1(3.33)	0	1(3.33)
49-54 months	11(34.38)	12(44.44)	3(9.38)	2(7.41)	2(6.25)	1(3.70)
5 year	13(28.26)	10(25.64)	4(8.70)	2(5.13)	3(6.52)	1(2.56)
6 year	7(18.92)	6(24.00)	4(10.81)	3(12.00)	3(8.11)	2(8.00)
7 year	7(17.95)	5(10.64)	7(17.95)	3(6.38)	3(7.69)	3(6.38)
8 year	5(8.77)	3(7.50)	3(5.26)	3(7.50)	2(3.51)	2(5.00)
9 year	7(16.28)	5(11.11)	3(6.98)	3(6.67)	2(4.65)	2(4.44)
10 year	4(12.12)	5(13.89)	3(9.09)	3(8.33)	3(9.09)	2(5.56)
11 year	6(12.24)	5(11.11)	2(4.08)	2(4.44)	1(2.04)	1(2.22)
12 year	4(9.52)	4(7.84)	2(4.76)	3(5.88)	2(4.76)	1(1.96)
13 year	3(8.57)	4(10.26)	2(5.71)	3(7.69)	0	1(2.56)
14 year	0	0	0	0	0	0

Figures in parentheses represent the percentage

form, it is 7 years and 6 years in boys and girls respectively. A severe form of caries is recorded highest at the 10 years (9.09%) and 6 years (8.00%) in boys and girls respectively.

Table 16.3: Prevalence of caries in permanent teeth by age and sex

Age (Year)	*Mild*		*Moderate*		*Severe*	
	Male	*Female*	*Male*	*Female*	*Male*	*Female*
6	0	0	0	0	0	0
7	0	0	0	0	0	0
8	0	0	0	0	0	0
9	1(2.33)	1(2.22)	1(2.33)	1(2.22)	0	0
10	2(6.06)	2(5.56)	0	1(2.78)	1(3.03)	1(2.78)
11	3(6.12)	3(6.67)	1(2.04)	0	0	0
12	2(4.76)	2(3.92)	0	0	0	0
13	3(8.57)	2(5.13)	2(5.71)	1(2.56)	3(8.57)	2(5.13)

(Cond...)

14	6(13.33)	7(13.21)	4(8.89)	3(5.66)	1(2.22)	1(1.89)
15	5(11.36)	4(10.53)	3(6.82)	3(7.89)	2(4.55)	2(5.26)
16	8(13.56)	6(8.96)	4(6.78)	5(7.46)	3(5.08)	2(2.98)
17	7(14.89)	8(14.29)	6(12.77)	7(12.50)	5(10.64)	4(7.14)
18	9(15.79)	10(14.93)	6(10.53)	5(7.46)	4(7.02)	4(5.97)
19	11(16.67)	8(14.55)	5(7.58)	5(9.09)	4(6.04)	3(5.45)
20	13(18.84)	11(16.92)	6(8.70)	7(10.77)	4(5.80)	3(4.62)
21	11(17.74)	14(18.92)	6(9.68)	9(12.16)	5(8.06)	6(8.11)
22	8(19.05)	1621.33)	5(11.90)	9(12.00)	3(7.14)	7(9.33)
23	12(19.05)	14(20.29)	7(11.11)	6(8.70)	5(7.94)	5(7.25)
24	8(17.78)	13(19.12)	4(8.89)	5(7.35)	3(6.67)	3(4.41)
25	9(20.00)	13(18.57)	6(13.33)	6(8.57)	3(6.67)	2(2.86)

Figures in parentheses represent the percentage.

Table 16.3 presents the numbers and percentages of caries affected permanent teeth by age and sex. Caries affected permanent teeth are absent at 6, 7, and 8 years of age. At the age of 9 years, mild and moderate forms of caries are observed in both males and females. The percentages of caries affected teeth increase with the increase of age. In males, the highest percentages of mild, moderate, and severe forms of caries are observed at the age of 25, 25, and 17 years respectively. In females, the highest frequency of mild was recorded at 23 years, moderate at 17 years, and severe at 22 years of age.

Table 16.4: Position and age group-wise distribution of caries affected permanent teeth in male

Age - group (year)	*No. of subjects*	*Anterior teeth*	*Percentage*	*Posterior teeth*	*Percentage*	*Anterior and posterior combined*	*Percentage*
6-10	206	2	0.97	5	2.43	7	3.40
11-15	215	14	6.51	29	13.49	43	20.00
16-20	298	35	11.74	72	24.16	107	35.91
21-25	257	46	17.90	84	32.68	130	50.58
26-30	217	47	21.66	94	43.32	141	64.98
31-35	197	44	22.34	86	43.65	130	65.99
36-40	142	33	23.24	62	43.66	95	66.90
41-45	117	24	20.51	52	44.44	76	64.95
46-50	117	27	23.08	55	47.01	82	70.09
51-55	79	19	24.05	32	40.51	51	64.56
56-60	98	24	24.49	41	41.84	65	66.33
61-65	51	13	25.49	22	43.14	35	68.63
66-70	30	8	26.67	13	43.33	21	70.00
70+	61	17	27.87	31	50.82	48	78.69
Total	2085	353	16.93	678	32.52	1031	49.45

The position and age-group wise distribution of caries affected permanent teeth among the Mizo male are shown in table 16.4. It is observed that the percentages of caries affected teeth increase with the increase of age. In all the age groups, the percentages of caries affected teeth are higher in the posterior teeth (premolars and molars) than the anterior teeth (incisors and canines). By 70 ≤ years, 50.82 percent of the posterior teeth are found affected by caries while only 27.87 percent of anterior teeth are affected.

Table 16.5: Position and age-group wise distribution of caries affected permanent teeth in female

Age group (year)	*No. of subjects*	*Anterior teeth*	*Percentage*	*Posterior teeth*	*Percentage*	*Anterior and posterior combined*	*Percentage*
6-10	193	1	0.52	5	2.59	6	3.11
11-15	226	9	3.98	36	15.93	45	19.91
16-20	310	36	11.61	88	28.39	124	40.00
21-25	356	70	19.66	128	35.96	198	55.62
26-30	281	54	19.22	132	40.93	186	60.15
31-35	205	42	20.49	74	43.41	116	63.90
36-40	179	37	20.67	81	45.25	118	65.92
41-45	156	36	23.08	56	46.15	92	69.23
46-50	134	31	23.13	45	44.03	76	67.16
51-55	112	27	24.11	33	46.43	60	70.54
56-60	92	25	27.17	19	45.65	44	72.82
61-65	68	19	27.94	21	48.53	40	76.47
66-70	32	9	28.13	9	53.13	18	81.26
70+	69	21	30.43	40	57.97	61	88.40
Total	2413	1121	17.28	2118	31.79	2959	49.07

Table 16.5 shows the position and age-group wise distribution of caries affected permanent teeth in females. The percentages of carious teeth increase with the increase of age. Except for 56-60 years, in all the age-groups, carious teeth in the posterior teeth appeared higher in frequency than the anterior teeth. In 70 ≤ years, the percentages of caries affected teeth are 30.43 percent and 57.97 percent in the anterior and posterior position of jaws respectively.

Table 16.6: Prevalence of caries according to literacy.

Literacy	Male			Female			Total		
	Present	Absent	Total	Present	Absent	Total	Present	Absent	Total
Non-literate	5	2	7	9	3	12	14	5	19
%	71.43	28.57	0.290	75.00	25.00	0.44	73.68	26.32	0.37
Literate	792	1610	2402	871	1859	2730	1677	3455	5132
%	32.97	67.03	99.71	31.90	68.10	99.56	32.68	67.32	99.63
Total	797	1612	2409	880	1862	2742	1691	3460	5151
%	33.08	69.92	100.00	32.09	67.91	100.00	32.83	67.17	100.00

Bisexual variation $\chi^2 = 0.79$, df = 1, $p > 0.05$
According to literacy in male $\chi^2 = 4.64$, df = 1, $p < 0.05$
According to literacy in female $\chi^2 = 10.18$, df = 1, $p < 0.05$

Table 16.6 shows the prevalence of caries by literacy among the Mizos. In both the males and females, the occurrences of dental caries are found much higher among the non-literate subjects than the literate ones. The variations in dental caries according to literacy is found highly significant in both the males ($\chi^2 = 4.64$, df = 1, $p < 0.05$) and females ($\chi^2 = 10.18$, df = 1, $p < 0.05$). The above table also shows that there is a higher prevalence of caries in males than their female counterparts, however, this difference is statistically not significant ($\chi^2 = 0.79$, df = 1, $p > 0.05$).

Table 16.7: Prevalence of caries according to income level

		Low-income group		Middle-income group		High-income group		Total
		Male	Female	Male	Female	Male	Female	
Unaffected	Number	507	562	705	811	400	489	3474
	%	61.53	59.28	72.23	71.33	65.68	74.43	67.44
Affected	Number	317	386	271	326	209	168	1677
	%	38.47	40.72	27.77	28.67	34.32	25.57	32.56
Grand total		824	948	976	1137	609	657	5151
%		15.99	18.4	18.95	22.07	11.82	12.75	100.00

Bisexual variation $\chi^2 = 7.29$, df = 2, $p < 0.05$
According to income in male $\chi^2 = 27.21$, df = 2, $p < 0.05$
According to income in female $\chi^2 = 51.29$, df = 2, $p < 0.05$

Table 16.7 shows the occurrence of dental caries among the Mizos according to their income level. The prevalence of dental caries was found highest in the low-income group. The occurrence of caries is inversely related to income. In both the low and middle-income groups, the percentages of caries are recorded higher in females than the males. The bisexual variation of the prevalence of caries according to the income is found statistically significant ($\chi^2 = 7.29$, df = 2, $p < 0.05$). The difference of the same, in respect of income among the males ($\chi^2 = 27.21$, df = 2, $p < 0.05$) and females ($\chi^2 = 51.29$, df = 2, $p < 0.05$) is found highly significant.

Table 16.8: Prevalence of caries among the betel nut (*Kuhva*) chewers and non-chewers

Sex		*Kuhva chewers*		*With caries*		*Non-chewers*		*With caries*	
		Number	*%*	*Number*	*%*	*Number*	*%*	*Number*	*%*
Male	2409	892	37.03	104	11.66	1517	62.97	693	45.68
Female	2742	783	28.56	102	13.03	1959	71.44	778	39.71
Total	5151	1573	30.54	206	4.00	3578	69.46	1471	28.56

Bisexual variation — $\chi^2 = 0.57$, df = 1, $p > 0.05$
Betel nut chewer and non-chewer males — $\chi^2 = 160.78$, df=1, $p<0.05$
Betel nut chewer and non-chewer females — $\chi^2 = 103.85$, df=1, $p < 0.05$

Table 16.8 shows the prevalence of caries among the betel nut (*kuhva*) chewers and non-chewers. Prevalence of caries found very low among the betel nut chewers than the non-chewers in both the sexes and the variation is statistically significant in both the males ($\chi^2 = 160.78$, df=1, $p<0.05$) and the females ($\chi^2 = 103.85$, df=1, $p < 0.05$), while no significant difference is observed in respect of the bisexual variation ($\chi^2 = 0.57$, df = 1, $p > 0.05$).

Table 16.9: Prevalence of caries among the smokers and non-smokers

Sex	*No. of subjects*	*Smokers*		*With caries*		*Non-smokers*		*With caries*	
		Number	*%*	*Number*	*%*	*Number*	*%*	*Number*	*%*
Male	2409	754	31.30	114	15.12	1655	68.70	217	13.11
Female	2742	456	16.63	76	16.67	2286	83.37	241	10.54
Total	5151	1049	20.36	190	18.11	4102	79.64	458	11.17

Bisexual variation among smokers — $\chi^2 = 0.37$, df = 1, $p> 0.05$
Bisexual variation among non-smokers — $\chi2 = 4.88$, df = 1, $p < 0.05$

The prevalence of caries among smokers and non-smokers is given in table 16.9. It shows that the frequency of caries is higher among smokers than non-smokers in both sexes. Among the males, 15.12 percent and 13.11 percent are caries affected smokers and non-smokers respectively and among the females, the percentages of the same are 16.67 percent and 10.54 percent respectively.

Table 16.10: Prevalence of caries among the tobacco chewers and non-chewers

Sex	No. of subjects	Tobacco chewers		With caries		Non-chewers		With caries	
		Number	%	Number	%	Number	%	Number	%
Male	2409	683	28.35	178	26.06	1726	71.65	368	21.32
Female	2742	948	34.57	253	26.69	1794	65.43	471	26.25
Total	5151	1631	31.66	431	26.43	3520	68.34	839	23.84

Bisexual variation among the chewers $\chi^2 = 0.05$, df = 1, $p > 0.05$
Bisexual variation among the non-chewers $\chi^2 = 7.27$, df = 1, $p < 0.05$
Tobacco chewer vs. non-chewer males $\chi^2 = 3.89$, df = 1, $p < 0.05$.
Tobacco chewer vs. non-chewer females $\chi^2 = 0.04$, df = 1, $p > 0.05$

Table 16.10 shows the prevalence of caries among tobacco chewers and non-chewers. In both the sexes, the prevalence of caries is found higher among the tobacco chewers than the non-chewers. Among both the tobacco chewers and non-chewers, the frequency of caries is recorded higher in females than their male counterparts. This Table also reveals that tobacco chewing is more common among females than males.

Conclusion

In the present study, both the deciduous and the permanent teeth were included. Teeth of 5151(M=2409; F=2742) Mizo individuals of Aizawl town, Mizoram belonging to all age groups were examined. 32.65 percent of the total population studied were affected by dental caries. Bisexual variation in respect of the prevalence of dental caries was not much (males, 48.0%; females 52.0%). Percentage of caries affected subjects found highest in the highest age group i.e., 70 years and above. The occurrence of dental caries was observed earliest at the age -group of 13-18 months in both the boys and girls where only mild and moderate forms were present. The frequency of caries affected teeth was recorded higher in the posterior teeth than the anterior teeth.

The highest prevalence of caries was observed among the low-income group in the present population. Income seems to have some correlation with the incidence of dental caries. Among the *kuhva* (betel nut and leaf) chewers (12.31%), the percentage of caries is less compared to the non-chewers (42.27%). Among the smokers; females (16.71%) are more prone to caries than their male (15.16%) counterparts. In respect of their tobacco chewing habit also, the females are in higher percentages than their male counterparts. The present study also reveals that the incidence of caries is much higher among the non-literates than the literates of both sexes. The findings of the present study reveal that chewing of *Kuhva* protects teeth from decay, but smoking and tobacco chewing increase the incidence of dental caries among the Mizo population. On the other hand, socio-economic factors like income and education also plays important role in controlling dental caries among the Mizos. As it was expected, the prevalence of dental caries was found inversely related to income.

References

Barret, JJ & Williamson, JJ 1972, 'Oral health of Australian aborigines: Survey methods and prevalence of dental caries', *Australian Dental Journal,* vol. 17, pp. 37-50.

Barnard, PD 1956, Dental survey of state school children in New South Wales. National Health and Medical Research Council. *Special Report Series* No. 8. Canberra.

Goldstein, MS 1948, 'Dentition of Indian crania from Texas', *American Journal of Physical Anthropology*, vol. 23, pp. 261-293.

Heitetz, SB, Horowitz, HS & Korts, DS 1976, 'Prevalence of dental caries in White and Black children in Nelson country, Virginia: A rural southern community', *Journal of Public Health Dentistry,* vol. 36, p. 79.

Hillson, S 1996, *Dental Anthropology*, Cambridge University Press.

James, MK 1979, *All About Dental Care*, Second Bell Books Edition. Vikash Publishing House Pvt. Ltd New Delhi.

Nag, CR 1993, *The Mizo Society in Transition*, Vikas Publishing House, New Delhi.

Nelson, CT 1938, 'The teeth of the Indians of Pecos Pueblo', *American Journal of Physical Anthropology*, vol. 23, pp. 261-293.

Pal, A. 1964, 'Observation on the dentition of crania from Eastern India', *Bulletin of the Anthropological Survey of India*, vol. 13, no. 1-2, pp. 9-18.

Rami Reddy, V & Naidu, GP 1980, 'Blood groups and dental caries in the Madigas of Chittoor district, Andhra Pradesh', *Indian Journal of Heredity,* vol. 12, pp. 81-91.

Rami Reddy, V, Vijayalakshmi, PB & Chandrasekhar Reddy, BK 1982b, 'Some dental features among the Muslims of South-eastern Andhra Pradesh', *Man in India,* vol. 62, no. 4, pp. 395-404.

Shourie, KL 1946, 'Eruption age of teeth in India', *Journal of Medical Research,* vol. 34, no. 1, pp. 105-118.

World Health Organization 1977, *Oral Health Surveys-Basic Methods* (2nd edition), Geneva.

Chapter - 17

Anthropometric Assessment in Identifying Risk Factors in Metabolic Syndrome

Moajungla Longkumer and B.T. Langstieh

Introduction

"Metabolic syndrome means a cluster of metabolic abnormalities leading to increased risk of cardiovascular disease, diabetes mellitus and other chronic diseases" (Chen *et al.,* 2015, p. 226) in simple words, it can be regarded as a seed that when planted will grow into a tree of chronic diseases. The identification of metabolic syndrome given as a joint interim statement of the International Diabetes Federation Task Force on Epidemiology and Prevention; National Heart, Lung, and Blood Institute; American Heart Association; World Heart Federation; International Atherosclerosis Society; and International Association for the Study of Obesity called as the Joint Scientific Statement (Alberti *et al.,* 2009) was adopted in the present study. This guideline stresses the ethnic difference in waist circumference and follows the international recommendation for thresholds of abdominal obesity as proposed by IDF (Alberti *et al.,* 2005) and is considered as one of the important components of metabolic syndrome. Besides waist circumference, the disorders of glucose metabolism, hypertension, dyslipidemia, and obesity were common sharing grounds to define metabolic syndrome with organizations such as the World Health Organization (Alberti and Zimmet, 1998), the National Cholesterol Education Program–Adult Treatment Panel III (Grundy *et al.,* 2005) and the International Diabetes Federation (Alberti *et al.,* 2005).

The prevalence of metabolic syndrome in both developed and developing countries shows a rising trend. It was found to increase from 32.9 percent in 2003-2004 to 34.7 percent in 2011-2012 in the United States (Aguilar *et al.,* 2015). Its prevalence is rapidly increasing in developing countries, leading to increased morbidity and mortality due to type 2 diabetes mellitus and cardiovascular disease (Misra & Khurana, 2008). Mortality due to cardiovascular disease though found to be decreasing in developed countries is on the rise in low and middle-income countries (Mendis & Banerjee, 2010) and about 70.0 percent of people with diabetes live in low and middle-income groups (Whiting *et al.,* 2010). South Asia is also known as the home of the largest population of people with metabolic syndrome. In India, about one-third of the urban population in large cities has metabolic syndrome (Pandit *et al.,* 2012). In the North-eastern states of India, the prevalence of overweight and obesity is about 19.6 percent and 6.1 percent in women and 16. 8.0 percemt and 5.1 percent in men as found in a study among the Meiteis of Manipur (Singh, 2013). The rural and urban difference in the prevalence of overweight among the Hmar of Manipur was found to be 42.93 percent and 57.26 percent respectively (Lalnuneng, 2019). As poor nutrition early in life (Bruce and Hanson, 2010) and overweight and obese adults are found to be a risk factor for developing metabolic syndrome (Bhal *et al.,* 2014), earlier studies have found the prevalence of overweight Ao adults especially in urban areas (Temsutula, 2010) and underweight Ao children (Longkumer, 2013).

An accepted common notion is that aging is a growing risk for disease and ultimately death (WHO, 2018), and earlier literature have found a higher prevalence of metabolic syndrome among the older age groups (Gupta *et al.,* 2004, Bansal *et al.,* 2017, Chinwale *et al., 2018*). However recent findings suggest its manifestation among middle-age groups as well (Bener, 2010; Khan *et al.,* 2018) which may suggest that there are many underlying risk factors responsible for the development of the metabolic syndrome. This paper in particular is an attempt to find out risk factors specific for the studied population group using anthropometric measurements and indices.

Materials and methods

It is a cross-sectional study conducted among the Ao Naga of Nagaland residing in rural areas of Mokokchung district and urban areas of Mokokchung, Kohima, and Dimapur town areas. It consists of 604 participants (male = 236, female = 368) belonging to the age of 18 to 70 years. Definition of metabolic syndrome given as Joint Scientific Statement (Alberti *et al.,* 2009) was used to determine the prevalence of metabolic syndrome in the population, according to which a person qualifies to have metabolic syndrome if; Waist circumference (WC) for Asian population ≥ 90 cm for men and ≥ 80 cm, blood pressure: ≥ 130/85 mmHg, hypertriglyceridemia: ≥150 mg/dl (1.69 mmol/L), low high-density lipoprotein:< 40 mg/dl (1.03 mmol/L) for men and <50 mg/dl (1.29 mmol/L) for women, and glucose level: ≥100 mg/dl (5.6 mmol/L). The participants provided written informed consent and the study protocol was approved by the Institutional Ethics Committee for research on human samples/participants, North Eastern Hill University, Shillong, Meghalaya.

Selected anthropometric measurement such as height, weight, mid-upper arm circumference (MUAC) and skinfold thickness at four regions (biceps, triceps, subscapular and suprailiac) was taken on the participants from all the areas. The standard technique of taking the anthropometric measurement was followed as described by Weiner & Lourie (1981) and Heyward & Wangner (2004). Anthropometric indices such as body mass index (BMI) were used to assess the nutritional status. Harpenden Skinfold Calliper was used for taking the skinfold thickness. A standard mercury sphygmomanometer and stethoscope were used to determine blood pressure. A blood glucose monitoring system manufactured by Code-free was used to evaluate the fasting glucose level and a standardized cholesterol test device manufactured by Mission was used to evaluate the high-density lipoprotein and triglyceride level.

A structured schedule was used to collect information on age, sex, education, income, and marital status. Education level was categorized as primary (Class 6 and below), under matric (Class 7-10), Post-Matric (Class 11- Graduation), and Graduate and above. Marital status was categorized as married, unmarried, widowed, and separated. Income was categorized as LIG (below Rs. 5000), MIG (Rs. 5000 – Rs. 9166), and HIG (above Rs. 9166).

Chi-square test was used for test of statistical significance for categorical variables, independent sample t-test was used for continuous variable. A correlation test was used to estimate the relationship between age and anthropometric measurements and simple logistic regression was done to find out the relationship between metabolic syndrome as the dependent variable and anthropometric measurements and age as an independent variable with the odds ratio (OR) and 95.0 percent confidence interval (CI). The analysis was performed using SPSS software.

Result

The demographic and socio-economic characteristics like age, education, the income of the studied population were displayed by subdividing the population under study into two basic categories: namely normal (without metabolic syndrome) and metabolic syndrome categories in rural and urban areas (Table 17.1).

Among the Ao Naga, the prevalence of metabolic syndrome was higher in urban areas (48.88%) than the rural counterparts (39.25%) and when male and female data were separated, in both rural (13.08%) and urban (16.89%) areas females had a higher prevalence of metabolic syndrome as compared to males (Figure 17.1). The influence of age on the prevalence of metabolic syndrome was also observed, it found a significantly higher prevalence in older age groups particularly among the 60 and above years in both rural (39.13%) and urban (28.95%) areas. Considering the education level concerning the prevalence of metabolic syndrome, it was found that, in rural areas, metabolic syndrome was more prevalent among individuals with primary (40.87%) and under matric (44.35%) educational levels but was not significant. In urban areas, metabolic syndrome was significantly more prevalent among individuals with under matric (55.26%) and post-matric (29.61%) educational levels ($p<0.01$). Again concerning income, in rural areas, the prevalence was significantly higher among LIG (53.04%) and MIG (26.96%) as compared to HIG (20.00%). Whereas, in urban areas, though not significant the prevalence was significantly higher among MIG (42.11%) and HIG (38.16%) as compared to LIG (19.74%).

Table 17.1: Demographic and socioeconomic characteristics of the study population by normal and metabolic syndrome categories.

	Rural (n=293)		***Urban (n=311)***	
	Normal N (%)	***MS N (%)***	***Normal N (%)***	***MS N (%)***
Number of individuals	178 (60.75)	115 (39.25)	159 (51.12)	152 (48.88)
Age group				
< 29 years	43 (24.16)	1 (0.87)	61 (38.36)	5 (3.29)
30-39	27 (15.17)	11 (9.57)	38 (23.90)	22 (14.47)
40-49	32 (17.98)	24 (20.87)	32 (20.13)	37 (24.34)
50-59	42 (23.60)	34 (29.57)	15 (9.43)	44 (28.95)
60 and above	34 (19.10)	45 (39.13)	13 (8.18)	44 (28.95)
	χ^2 = 38.582, df= 4, p= 0.000		χ^2 = 83.143, df= 4, p= 0.000	
Educational level				
Primary	52 (29.21)	47 (40.87)	13 (8.18)	18 (11.84)
Under matric	91 (51.12)	51 (44.35)	57 (35.85)	84 (55.26)
Post matric	30 (16.85)	17 (14.78)	68 (42.77)	45 (29.61)
Graduate and above	5 (2.81)	none	21 (13.21)	5 (3.29)
	χ^2 =6.888, df = 3, p= 0.076		χ^2 = 20.357, df= 3, p= 0.000	
Income groups				
LIG	126 (70.79)	61 (53.04)	40 (25.16)	30 (19.74)
MIG	28 (15.73)	31 (26.96)	74 (46.54)	64 (42.11)
HIG	24 (13.48)	23 (20.00)	45 (28.30)	58 (38.16)
	χ^2 =9.668, df= 2, p= 0.008		χ^2 = 3.638, df= 2, p= 0.162	

Table 17.2: Prevalence of metabolic syndrome in rural and urban areas by categories of marital status in males and females.

Marital Status	*Rural (n=293)*		*Urban (n=311)*	
	Normal N (%)	*MS N (%)*	*Normal N (%)*	*MS N (%)*
Male (n=236)				
Married	49 (60.49)	34 (94.44)	38 (55.07)	47 (94.00)
Unmarried	28 (34.57)	2 (5.56)	30 (43.48)	2 (4.00)
Widow & Separated	4 (4.94)	-	1 (1.45)	1 (2.00)
	$\chi^2 = 14.009$, df= 2, p =0.001		$\chi^2 = 23.006$, df = 2, p = 0.000	
Female (n=368)				
Married	61 (62.89)	61 (77.22)	41 (45.56)	76 (74.51)
Unmarried	24 (24.74)	9 (11.39)	43 (47.78)	10 (9.80)
Widow & Separated	12 (12.37)	9 (11.39)	6 (6.67)	16 (15.69)
	$\chi^2 = 5.463$, df= 2, p = 0.147		$\chi^2 = 34.949$, df = 2, p =0.000	

To study the prevalence pattern of metabolic syndrome according to marital status, male and female data were separated as given in table 17.2. Firstly, among males, in both rural and urban populations, the prevalence of metabolic syndrome was significantly higher among individuals who are married (94.44% and 94.00%, respectively), while unmarried males were found with less prevalence of metabolic syndrome ($p<0.01$). Secondly, among females, in both rural and urban populations, though not significant, the prevalence of metabolic syndrome was higher among individuals who are married (77.22% and 74.51%, respectively).

It is evident from the above tables that there are differences in the prevalence of metabolic syndrome between different demographic and socio-economic conditions and there is the possibility that anthropometric characters between normal (without metabolic syndrome) and metabolic syndrome will be different as such an attempt has been made to show in table 17.3(A) and 17.3(B) for males and females, respectively.

Table 17.3(A): Anthropometric measurements of rural and urban males

Variables	*Rural (n=117)*			*Urban (n=119)*		
	Normal Mean (SD)	*MS Mean (SD)*	*t-test*	*Normal Mean (SD)*	*MS Mean (SD)*	*t-test*
Height	162.67 (6.59)	163.41 (4.78)	0.601	163.79 (5.38)	163.08 (4.87)	0.740
Weight	57.74 (7.99)	71.10 (10.47)	7.556**	60.39 (9.49)	71.01 (8.75)	6.226**
Hip circumference	90.48 (7.01)	99.77 (5.75)	6.965**	92.08 (7.71)	100.20 (5.42)	6.638**
Waist circumference	79.38 (7.92)	92.33 (7.69)	8.228**	81.29 (9.27)	93.99 (7.09)	8.116**

(Cond...)

WHR	0.87 (0.03)	0.92 (0.04)	7.329**	0.88 (0.03)	0.93 (0.04)	7.939**
MUAC	26.67 (2.49)	29.88 (2.74)	6.221**	26.72 (2.59)	29.24 (2.29)	0.548**
SF bicep	4.05 (1.07)	6.16 (1.79)	7.897**	4.60 (1.60)	6.29 (1.95)	5.169**
SF triceps	6.60 (2.01)	9.47 (2.78)	6.313**	7.98 (2.79)	10.12 (2.91)	4.058**
SF subscapular	11.40 (2.85)	17.81 (4.49)	9.301**	12.77 (4.16)	17.37 (3.70)	6.223**
SF suprailiac	7.10 (2.32)	11.12 (4.04)	6.793**	9.61 (3.84)	12.06 (3.54)	3.554**
BMI	21.78 (2.33)	26.59 (3.34)	8.968**	22.55 (3.27)	26.67 (2.63)	7.337**

P<0.01**, p<0.05*

Table 17.3(B): Anthropometric measurements of rural and urban females

Variables	*Rural (n=176)*			*Urban (n=192)*		
	Normal Mean (SD)	*MS Mean (SD)*	*t-test*	*Normal Mean (SD)*	*MS Mean (SD)*	*t-test*
Anthropometric measurement						
Height	152.71 (7.08)	152.51 (4.96)	0.207	155.36 (5.06)	154.25 (5.06)	1.507
Weight	49.79 (6.28)	59.34 (8.21)	8.742**	53.04 (9.68)	62.75 (7.93)	7.636**
Hip circumference	93.42 (6.66)	102.35 (6.53)	8.924**	93.98 (7.96)	103. 36 (6.56)	8.949**
Waist circumference	79.15 (7.04)	91.57 (8.13)	10.854**	80.29 (9.70)	93.49 (8.27)	10.176**
WHR	0.85 (0.02)	0.89 (0.04)	9.498**	0.85 (0.03)	0.90 (0.04)	9.715**
MUAC	24.86 (1.92)	28.46 (2.55)	10.669**	25.01 (2.76)	27.96 (2.30)	7.967**
SF bicep	6.20 (2.09)	8.47 (2.42)	6.680**	7.28 (2.54)	9.33 (2.78)	5.300**
SF triceps	12.74 (3.20)	16.47 (4.33)	6.554**	14.41 (4.16)	17.80 (4.60)	5.321**
SF subscapular	14.65 (13.28)	18.55 (5.32)	2.450**	14.77 (5.12)	15.12 (5.23)	5.416**
SF suprailiac	11.62 (4.25)	16.02 (5.42)	6.031**	13.02 (5.50)	17.29 (3.89)	6.277**
BMI	21.18 (2.27)	25.53 (3.12)	10.682**	21.81 (3.23)	26.37 (3.003)	10.130**

P<0.01**, p<0.05*

In males and females, from both rural and urban areas, the anthropometric measurements such as weight, waist circumference, hip circumference, mid-upper arm circumference, skinfold measurements for bicep, triceps, subscapular, and suprailiac including body mass index (BMI) were found to be significantly higher for individuals with metabolic syndrome as compared to normal.

The relationship of anthropometric measurement with age was evaluated using correlation analysis performed among normal and metabolic syndrome individuals is shown in table 17.4.

Table 17.4: Correlation test between age and anthropometric measurements in normal and metabolic syndrome individuals

Anthropometric measurements	*Male (n=236)*				*Female (n=368)*			
	Rural (n=117)		*Urban (n=119)*		*Rural (n=176)*		*Urban (n=192)*	
	Normal	*MS*	*Normal*	*MS*	*Normal*	*MS*	*Normal*	*MS*
Height	0.034	-0.206	-0.437**	-0.304*	-0.355**	-0.260*	-0.167	-0.371 **
Weight	0.194	-0.356*	0.029	-0.057	-0.147	-0.284*	0.096	-0.123
Hip circumference	0.495**	0.010	0.392**	0.087	0.085	-0.001	0.363**	0.136
Waist circumference	0.495**	-0.096	0.377**	0.213	0.100	-0.023	0.351*	0.184
MUAC	0.160	-0.363*	0.055	-0.256	0.020	-0.162	0.276**	-0.014
SF bicep	0.260*	-0.253	0.132	-0.229	-0.397**	-0.221*	-0.065	-0.006
SF triceps	0.361**	-0.278	0.102	-0.373**	-0.040	-0.013	0.069	0.110
SF subscapular	0.181	-0.027	0.189	-0.098	-0.134	-0.282	0.016	-0.213*
SF suprailiac	0.236*	-0.096	0.233	0.002	-0.004	-0.043	0.168	0.024
BMI	0.230*	-0.326	0.213	0.106	0.084	-0.172	0.146	0.082

P<0.01**, p<0.05*

Interestingly, an inverse association was found among individuals with metabolic syndrome, this may be an indication of higher anthropometric measurements found among metabolic syndrome despite their age being young or in other words the prevalence of metabolic syndrome even among younger individuals may be because of their higher anthropometric measurements. In males with metabolic syndrome, significant inverse associations were found in weight ($p<0.05$), and MUAC ($p<0.05$) among the rural population and in height ($p<0.05$) and skinfold at triceps ($p<0.01$) among the urban population. In females with metabolic syndrome, a significant inverse association was found in height ($p<0.05$), weight ($p<0.05$), and skinfold at bicep ($p<0.05$) among the rural population and in height ($p<0.05$) and skinfold at subscapular ($p<0.05$) among the urban population.

Table 17.5: Logistic regression (odds ratio) for metabolic syndrome by anthropometric measurements and age by the categories of rural and urban male

	Rural (n=117)				*Urban (n=119)*			
	Sig.	*OR*	*95% of CI*		*Sig.*	*OR*	*95% of CI*	
			Lower	*Upper*			*Lower*	*Upper*
Age	p<0.05	1.036	1.006	1.068	p<0.01	1.073	1.041	1.106
Height	p>0.05	1.020	.956	1.089	p>0.05	.973	.906	1.045
Weight	p<0.01	1.175	1.104	1.250	p<0.01	1.140	1.080	1.203
Hip circumference	p<0.01	1.245	1.141	1.359	p<0.01	1.198	1.112	1.291
Waist circumference	p<0.01	1.233	1.138	1.336	p<0.01	1.204	1.123	1.290
MUAC	p<0.01	1.586	1.312	1.918	p<0.01	1.506	1.257	1.805
SF bicep	p<0.01	2.625	1.827	3.771	p<0.01	1.732	1.342	2.235
SF triceps	p<0.01	1.663	1.341	2.062	p<0.01	1.312	1.131	1.522
SF subscapular	p<0.01	1.598	1.346	1.897	p<0.01	1.329	1.185	1.490
SF suprailiac	p<0.01	1.517	1.279	1.799	p<0.01	1.190	1.072	1.322
BMI	p<0.01	1.778	1.443	2.192	p<0.01	1.544	1.313	1.816

Table 17.6: Logistic regression (odds ratio) for metabolic syndrome by anthropometric measurements and age by the categories of rural and urban female

	Rural (n=176)				*Urban (n=192)*			
	Sig.	*OR*	*95% of CI*		*Sig.*	*OR*	*95% of CI*	
			Lower	*Upper*			*Lower*	*Upper*
Age	p<0.01	1.069	1.041	1.098	p<0.01	1.121	1.084	1.158
Height	p>0.05	0.995	0.948	1.044	p>0.05	.957	.904	1.014
Weight	p<0.01	1.213	1.141	1.291	p<0.01	1.140	1.092	1.190
Hip circumference	p<0.01	1.231	1.153	1.314	p<0.01	1.193	1.130	1.259
Waist circumference	p<0.01	1.250	1.170	1.335	p<0.01	1.176	1.122	1.232
MUAC	p<0.01	2.396	1.844	3.113	p<0.01	1.585	1.367	1.839
SF bicep	p<0.01	1.557	1.328	1.826	p<0.01	1.341	1.186	1.515
SF triceps	p<0.01	1.306	1.183	1.441	p<0.01	1.190	1.107	1.278
SF subscapular	p<0.05	1.064	1.007	1.125	p<0.01	1.175	1.099	1.256
SF suprailiac	p<0.01	1.204	1.120	1.294	p<0.01	1.216	1.132	1.306
BMI	p<0.01	2.052	1.651	2.551	p<0.01	1.612	1.406	1.847

To further investigate the association of anthropometric measurements with metabolic syndrome logistic regression was performed, with an attempt to find out the risk factors and estimate the odds ratio with 95.0 percent CI in males (Table 17.5) and females (Table 17.6). Among males, in both rural and urban areas, the odds of developing metabolic syndrome significantly increase with the increase in age, weight, hip circumference, waist circumference, MUAC, skinfold measurements, and BMI. Similarly, among females, in both rural and urban areas, the odds of developing metabolic syndrome significantly increase with the increase in age, weight, hip circumference, waist circumference, MUAC, skinfold measurements, and BMI.

Discussion

"The prevalence of metabolic syndrome varies depending on the definition applied, ethnicity and the age of the subjects in the studied population" (Kaur, 2014, p. 141). In the present study, the overall prevalence of metabolic syndrome in the population was more in the urban areas (48.88%) as compared to rural counterparts (39.25%). Much earlier literature on epidemiological studies have similarly reported a higher prevalence of metabolic syndrome in urban and semi-urban population when compared to rural populations (Weng *et al.,* 2006; Ulasi *et al.,* 2010; Gupta *et al.,* 2004; Mishra *et al.,* 2006; Prabhakaran *et al.,* 2007; Mahadik *et al.,* 2007). However, some studies have found highly prevalent in rural population just as in urban population (Aekplakron *et al.,* 2009) concluding that metabolic syndrome is not only confined to the urban population but also rural populations as well.

Increasing age is a risk factor found in both urban and rural populations (Aekplakron *et al.,* 2009, Aguilar *et al.,* 2015, Lokanath *et al.,* 2014) and the present study have also confirmed increase in age to be a contributing factor for the development of metabolic syndrome in populations under study.

The relationship between some anthropometric measurements as a risk factor and the development of metabolic syndrome became quite clear with the estimation of the odds ratio. The study found the odds of developing metabolic syndrome significantly increase with the increase in age, weight, hip circumference, waist circumference, MUAC, skinfold measurements, and BMI in males and females living in both rural and urban areas. Previous research studies have also found that the value of BMI is higher among individuals with metabolic syndrome in other countries (Mahadik *et al.,* 2007; Aekplakorn *et al.,* 2009; Yu *et al.,* 2016) and in India (Lokanath *et al.,* 2014). According to Misra and Khurana (2008), the development of metabolic syndrome and cardiovascular risk in South Asians is heightened by higher body fat, truncal subcutaneous fat, intra-abdominal fat, and ectopic fat deposition (liver fat, etc.). Abdominal obesity was reported to be a driving component for the occurrences of metabolic syndrome (Ghosh 2005, Adejamo *et al.,* 2019) because abdominal obesity causes the overproduction of adipocytokines. It integrates with endocrine, autocrine, and pancreatic signals and affects the multiple processes of insulin sensitivity, oxidant stress, energy metabolism, and blood coagulation, and inflammatory response which accelerates atherosclerosis, plaque rupture, and atherothrombosis (Kaur, 2014 b). Another study found body roundness index and waist circumference were associated significantly with metabolic syndrome and insulin resistance (Li *et al.,* 2019). However, some studies found its prevalence even in non-obese individuals (Mahadik *et al.,* 2007) where the central fat deposition was found to be a risk factor (Momesso *et al.,* 2011).

Conclusion

Though the prevalence of metabolic syndrome was found to be slightly higher in the urban population it was not, however, absent from this rural population in North-east India. This aligns with the fact that metabolic syndrome is known as the outcome of overnutrition, energy surplus, and physical inactivity, increasingly being seen in countries of the developing world and now undergoing the transition to more urban. Increasing age and anthropometric measurements were found to be risk factors for the prevalence of metabolic syndrome in the studied population.

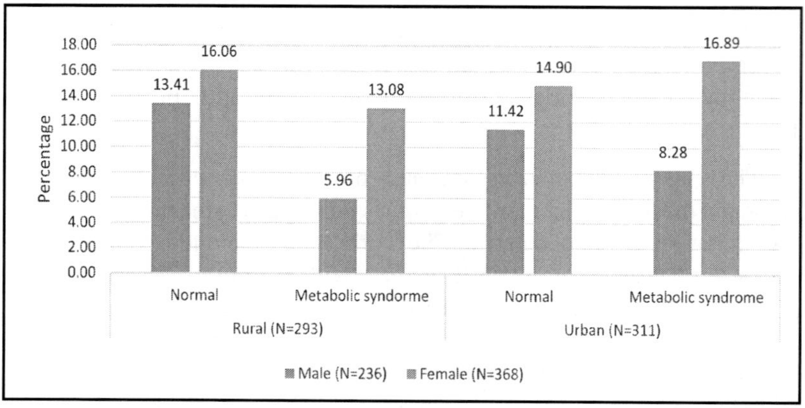

Figure 17.1: Prevalence of metabolic syndrome in rural and urban population by categories of sex.

References

Adejumo, EN, Adejumo, AO, Azenabor A, Ekun, AO, Enitan, SS, Adebola, OK and Ogundahunsi, OA 2019, 'Anthropometric parameters that best predict metabolic syndrome in Southwest Nigeria'. *Diabetes and Metabolic Syndrome: Clinical Research and Reviews*. Vol. 13, pp. 48-54.

Aekplakorn, W, Kessomboon, P, Sangthong, R, Chariyalertsak, S, Putwatana, P, Inthawong, R, Nitiyanant, W and Taneepanichskul, S 2009, 'Urban and rural variation in clustering of metabolic syndrome components in the Thai population: Results from the fourth National Health Examination Survey'. *Bio-Med Central Public Health*. Vol. 11, no. 854, pp. 1-9.

Aguilar, M, Bhuket, T, Torres, S, Liu, B, Robert, J and Wong, RJ 2015, 'Prevalence of the metabolic syndrome in the United States, 2003-2012'. *Journal of the American Medical Association*. Vol. 313, no. 19, pp. 1973-1974.

Alberti, KG, and Zimmet, PZ 1998, 'Definition, diagnosis and classification of diabetes mellitus and its complications, part 1: Diagnosis and classification of diabetes mellitus provisional report of a WHO consultation'. *Diabetic Medicine*, Vol. 15, no. 7, pp. 539-553.

Alberti, KG, Zimmet, P, Shaw, J 2005, 'IDF Epidemiology Task Force Consensus Group. The metabolic syndrome: a new worldwide definition. *Lancet*. Vol. 366, pp. 1059-1062.

Alberti, KGMM, Eckel, RH, Grundy, SM, Zimmet, PZ, Cleeman, JI, Donato, KA, Fruchart, JC, James, WPT, Loria, CM, Smith, SC 2009, 'Harmonizing the metabolic syndrome, A Joint Interim Statement of the International Diabetes Federation Task Force on Epidemiology and Prevention; National Heart, Lung, and Blood Institute; American Heart Association; World Heart Federation; International Atherosclerosis Society; and International Association for the Study of Obesity'. *Circulation*. Vol. 120, pp. 1640-1645.

Bahl, D, Singh, K, Sabharwal, M and Arora, M 2014, 'Anthropometric indices for the prediction of metabolic syndrome and its features, among children and adolescents'. *Indian Journal of Science and Technology*. Vol 7, no.8, pp. 1066-1077.

Bansal, S, Paliwal, A, Verma, V and Chauhan, J 2017, 'A study on prevalence of metabolic syndrome in general population in Western Uttar Pradesh, India'. *International Journal of Research in Medical Sciences*. Vol. 5, no. 6, pp. 2641-2643.

Bener, A, Mohammad, A, Ismail, AN, Zirie, M, Abdullatef, WK and Al-Hamaq, AOAA 2010, 'Gender and age-related differences in patience with metabolic syndrome in a highly endogamous population'. *Bosnian Journal of Basic Medical Sciences*. Vol. 10, no. 3, pp. 210-217.

Bruce, KD and Hanson, MA 2010, 'Developmental origins, mechanisms and implications of metabolic syndrome'. *Journal of Nutrition*. Vol. 140, no. 3, pp. 648-652. Doi: 10.3945/jn.109.111179.

Chen, LY, Zhu, WH, Chen, ZW, Dai, HL, Ren, JJ, Chen, JH, Chen, LQ and Fang, LZ 2007, 'Relationship between hyperuricemia and metabolic syndrome'. *Journal of Zhenjiang University Science B*. Vol. 8, no. 8, pp. 593-598.

Chinawale, CG, Parmar, DV, Kavathia P, Rangnani, T, Thakkar, J and Kartha, G 2018, 'Metabolic syndrome among adults of Surendranagar District of Saurashtra, Gujarat: A cross-sectional study'. *Indian Journal of Community Medicine*. Vol. 43, pp. 24-28.

Ghosh, A. 2005, 'Factor analysis of metabolic syndrome among the middle-aged Bengalee Hindu men of Calcutta, India'. *Diabetes/Metabolism Research and Review*. Vol. 21, no.1, pp. 58-64.

Grundy, SM, Cleeman, JI, Daniels, SR, Donato, KA, Eckel, RH, Franklin, BA, Gordon, DJ, Krauss, RM, Savage, PJ, Smith, SC Jr., Spertus, JA, Costa, F, American Heart Association; National Heart, Lung, and Blood Institute 2005, 'Diagnosis and management of the metabolic syndrome: An American Heart Association/National Heart, Lung, and Blood Institute Scientific Statement'. *Circulation*, Vol. 112, no. 17, pp. 2735–2752. Doi: 10.1161/CIRCULATIONAHA.105.169404.

Gupta, R, Deedwania, PC, Gupta, A, Rastogi, S, Panwar, RB and Kothari, K 2004, 'Prevalence of metabolic syndrome in an Indian urban population'. *International Journal of Cardiology*. Vol. 97, no. 2, pp. 257-261.

Heyward, VH and Wanger, DR 2004, '*Applied Body Composition Assessment*', 2nd edition. United States: Human Kinetics.

Kaur, J 2014 a, 'Assessment and screening of the risk factors in metabolic syndrome'. *Medical Sciences*, Vol. 2, pp. 140-152. Doi:10.3390/medsci2030140.

Kaur, J 2014 b, 'A Comprehensive review on metabolic syndrome'. *Cardiology Research and Practice*, *2014*, pp. 943162. Doi: org/10.1155/2014/943162.

Khan, Y, Lalchandani, A, Gupta, AC, Khadanga, S and Kumar, S 2018, 'Prevalence of metabolic syndrome crossing 40% in Northern India: Time to act fast before it runs out of proportions'. *Journal of Family Medicine and Primary Care*. Vol. 7, pp. 118-123.

Lalnuneng, A. 2019, '*A study of blood pressure in relation to body composition and socioeconomic status among the Hmars of Manipur*. Ph.D. thesis, North Eastern Hill University.

Li, G, Wu, H, Wu, X, Cao, Z, Tu, Y, Ma, Y, Li, B, Peng, Q, Cheng, J, Wu, B and Zhou, Z 2019, 'The feasibility of two anthropometric indices to identify metabolic syndrome, insulin resistance and inflammatory factors in obese and overweight adults'. *Nutrition*. Vol. 57, pp. 194-201.

Lokanath, DA, Chandrashekariah, SA, Xaviour, D and Rao, J 2014, 'The incidence and alliance of metabolic syndrome with cardiovascular risk markers among Kodavas'. *Open Journal of Endocrine and Metabolic Diseases*. Vol. 4, pp. 158-166.

Longkumer, T 2013, 'A study on somatotype and nutritional status in relation to physical activities among the Ao children of Mokokchung district, Nagaland'. Ph.D. thesis, North Eastern Hill University.

Mahadik, SR, Deo, SS and Mentalia, SD 2007, 'Increased prevalence of metabolic syndrome in non-obese Asian Indian- An urban-rural comparison'. *Metabolic Syndrome and Related Disorders*. Vol. 5, pp. 142-152.

Mendis, S and Banerjee, A. 2010, 'Cardiovascular disease, equity, and social determinants. In Blas, E. and Kurup, A.S. (Eds.). *Equity, social determinants and public health programs*'. Geneva: WHO press. pp. 31-48.

Misra, A and Khurana, L 2008, 'Obesity and the metabolic syndrome in developing countries. *Journal of Clinical Endocrinology Metabolism*. Vol. 93, no. 11, pp. 9-30.

Misra, KB, Endemann, SW and Ayer, M 2006, 'Measure of obesity and metabolic syndrome in Indian American in Northern California'. *Ethnicity and Disease*. Vol. 16, no. 2, pp. 331-337.

Momesso, DP, Bussade, I, Giovanna, A, Lima, B, Fonseca, LPC, Russo, LAT and Kupfer, R 2011, 'Body composition, metabolic syndrome and insulin resistance in type 1 diabetes mellitus'. *Arquivos Brasileiros De Endocrinologia & metabologia*. Vol. 55, no. 3, pp. 189-193.

Pandit, K, Goswami, S, Ghosh, S, Mukhopadhyay, P and Chowdhury, S 2012, 'Metabolic syndrome in South Asians'. *Indian Journal Endocrinology and Metabolism*. Vol. 16, pp. 44-55.

Prabhakaran, D, Chaturvedi, V, Pankaj, S, Manhapra, A, Panniyammakal, J, Shah, B, and Srinath, RK 2007, 'Difference in the prevalence of metabolic syndrome in urban and rural India: A problem of urbanization'. *Chronic Illness*. Vol. 3, no.1, pp. 8-10.

Singh, MS 2013, '*Obesity and self-reported morbidity in relation of lifestyle and diet among the Meitei of Manipur*'. Ph.D. thesis, North Eastern Hill University.

Temsutula, 2010 '*Body composition and nutritional status among the Ao adults*'. Ph.D. thesis, North Eastern Hill University.

Ulasi, I, Oljoma, CK, and Onodugo, OD 2010, 'A community-based study of hypertension and cardio-metabolic syndrome in semi-urban and rural communities in Nigeria'. *Bio-Med Central Health Service Research*. Vol. 10, no. 71, pp. 1-6.

Weiner, JS, and Lourie, JA 1981, *Practical human biology*. New York: Academic Press.

Weng, X, Liu, Y, Ma, J, Wang. W, Yang, G, and Caballero, B 2006, 'An urban-rural comparison of the prevalence of the metabolic syndrome in Eastern China'. *Public Health Nutrition*. Vol. 10, no. 2, pp. 131-136.

Whiting, D, Unwin, N and Roglic, G 2010, 'Diabetes, equity, and social determinants. In Blas, E. and Kurup, A.S (Eds.), *Social Determinants and Public health programmes*'. Geneva: WHO Press, pp. 77-94.

WHO 2018, 'Aging and health'. [online]. Available at www.who.int. [Accessed date: 13/09/2019].

Yu, TY, Jee, JH, Bae, JC, Jin, SM, Baek, JH, Lee, MK and Kim, JH 2016, 'Serum uric acid: A strong and independent predictor of metabolic syndrome after adjusting for body composition'. *Metabolism Clinical and Experimental*. Vol. 65, pp. 432-440.

Chapter - 18

Age Changes in Somatotypes among the Chakhesang Naga Boys of Dimapur District, Nagaland

VIVOTO TSUKRU AND JOHNSLY WELL DKHAR

Introduction

'Body physique' is a generic word for the whole body which refers to the shape and size of an individual. Somatotype is the quantification of human body shape and body composition, simplified in the form of a three-number rating scale, useful in describing the human physique. These three-number ratings represent endomorphy (relative fatness), mesomorphy (relative musculoskeletal robustness), and ectomorphy (relative linearity) always in the same order (Carter, 2002). Somatotyping is used as a measure of assessing body shape and composition, independent of size (Bailey *et al.,* 1982), and is widely applied in the field of growth, nutrition, sports, occupation, diseases, *etc.* (Carter and Heath, 1990). Anthropometric somatotype has been most widely applied in the field of sports, likewise, it has broad application in a variety of ways for assessing the effects of nutrition, environmental factors, and socio-economic variables on the body physique of both children and adults.

Adolescence is the period of rapid changes in the biology of human beings, and therefore, the changes in body physique during this stage are quite apparent. During the growth stage, somatotype variation is markedly observed between individuals of different age groups belonging to the same population (Kaul *et al.,* 1996). Studies show the age effect on somatotype shows more variance among adolescents than adults (Heath and Carter, 1971; Parizkova and Carter, 1976; Singh and Sidhu, 1980). Multiple studies reported age changes in somatotype of different Indian populations among adolescents (Bhasin and Jain, 2007; Gakhar and Malik, 2002; Ghosh and Malik, 2004; Handa *et al.,* 1995; Kaur and Pathak, 2014) and among adults (Kaur, 2009; Singh and Sidhu, 1980; Singh *et al.,* 2007; Ghosh and Malik, 2010; Thapliyal and Kumar, 2019).

There is quite a reasonable number of research works on the adolescent age effect on somatotype of the different Indian populations concerning varied geographical or altitudinal location, place of residence, socio-economic factors, caste groups, tribal groups, physical activities, and so on. Still, data on the physique characteristics of the Indian population is limited and only a handful of reports are available (Bhasin and Jain, 2007). The North-east Indian population is no exception in this case. Even Singh (2011) opines that there remains a dearth in the scientific literature on the somatotypes of different populations of North-east India. Research publications on somatotypes of adolescents in this part of India include- Gaur *et al.,* (1999) among Meitei children; Dkhar (2005) among Pnar boys; Longkumer (2014) among Ao Naga boys; Sarkar and Sil (2014) among Tripuri tribal boys; Sarkar and Sil (2014) among non-athlete tribal boys of Tripura; Sarkar and Sil (2015) among Bengali boys; Singh (2011) among Meitei boys.

The present chapter aims to describe the somatotype patterns of Chakhesang Naga boys aged between 8-18 years and to compare the findings with other reported studies on similar age groups of the North-east population.

Materials and methods

The present study was conducted in Sodzulhou village of Dimapur district, Nagaland in 2011. This village is inhabited by the Chakhesang Naga people. The word 'Sodzulhou' is the amalgamation of two words 'Sodzu' meaning 'far distance' and 'Lhou' meaning 'dwell'. Interestingly people from this village migrated to the present location of the village from Khezhakeno village under Phek district and settled in the outskirts of Dimapur district.

Anthropometric measurements were taken from a cross-sectional sample of 200 Chakhesang Naga boys aged between 8 to 18 years. Heath-Carter method of anthropometric somatotype (Heath and Carter, 1967) was employed for computing the somatotype ratings. This method incorporates the use of ten anthropometric measurements *viz.,* height, weight, bi-condylar humerus, bi-condylar femur, biceps girth, calf girth, skinfold at triceps, skinfold at sub-scapula, skinfold at supra-iliac, and skinfold at the calf.

These measurements were taken using standard techniques given by Weiner and Lourie (1981) on the left side of the subject. Care was taken to include only those subjects who were normal and healthy, free from any congenital disorders. The age groups are based on decimal age following the Decimal Age Calendar given by Tanner (Weiner and Lourie, 1981). Statistical analysis includes mean somatotypes, standard deviation, somatotype attitudinal mean (SAM) were calculated and student's t-test was used for comparative purpose.

Results and discussion

Table 18.1 presents the height, weight, height-weight ratio (HWR), somatotype components, and somatotype attitudinal mean (SAM) of Chakhesang Naga boys. The mean height, weight, and HWR at age 8 years is 118.7 cm, 25.52 kg, and 40.32 respectively, and at age 18 years is 164.02 cm, 52.72 kg, and 43.74 respectively. From age 8 to 18 years, the total increment in height and weight is 45.32 cm and 27.2 kg respectively. The maximum increment in height was observed between age groups 14 and 15 years. However, almost the same reading of increment was recorded between 13 and 14 years of age, indicating that height spurt occurs from 13 to 15 years of age among the studied population. For weight, the maximum increment was recorded between the age of 13 and 14 years. There is a regular pattern of increase in height and weight with age. However, HWR shows a slightly irregular pattern of increase with increasing age; but on the whole, it shows a general increasing trend from 40.32 at age 8 years to 43.74 at age 18 years. The somatotype attitudinal means (SAM) also shows an irregular pattern of increase and decrease from age 8 years to 18 years.

The mean somatotype of Chakhesang Naga boys at age 8 years is 1.44–4.07–1.92 and at age 18 years is 1.65–3.42–3.71. During these ten years, there was an overall increase of 0.20 units in endomorphy, 1.79 units in ectomorphy, and an overall decrease of 0.65 units in mesomorphy. The maximum and minimum somatotype rating of endomorphy was recorded at age 17 years and age 9 years respectively. For mesomorphy, maximum and minimum ratings were observed at age 9 years and age 15 years respectively. For ectomorphy, maximum somatotype rating was observed at age 16 years and minimum at age 9 years. In all the three somatotype components, there is a fluctuating increase and decrease from age 8 to 18 years with different degrees of variation (Figure. 18.1). This fluctuation is most prominent in ectomorphy and least prominent in the endomorphy component.

Table 18.1: Height, weight, and somatotype components of Chakhesang Naga boys by age

Age group (in years)	*N*	*Height*	*Weight*	*HWR*	*Endo-morphy*	*Meso-morphy*	*Ecto-morphy*	*SAM*
8	19	118.7	25.52	40.32	1.44 ± 0.32	4.07 ± 0.44	1.92 ± 1.26	1.49
9	18	125.07	31.22	39.72	1.36 ± 0.46	4.55 ± 0.64	1.47 ± 1.11	1.44
10	18	129.01	31.38	40.90	1.52 ± 0.26	4.16 ± 0.87	1.88 ± 1.38	1.77
11	18	134.6	32.72	42.08	1.55 ± 0.29	4.19 ± 0.90	2.75 ± 1.39	1.80
12	18	136.36	33.55	42.27	1.75 ± 0.54	4.02 ± 0.73	2.80 ± 1.16	1.89
13	18	139.41	35.94	42.24	1.50 ± 0.41	3.41 ± 0.82	2.38 ± 1.15	1.90
14	18	148.43	42.66	42.47	1.66 ± 0.37	3.52 ± 0.98	2.63 ± 1.19	1.80
15	18	157.48	46.44	43.81	1.61 ± 0.62	2.97 ± 0.81	3.55 ± 1.15	1.79
16	18	159.22	47.88	43.85	1.61 ± 0.43	3.08 ± 0.86	3.77 ± 1.16	2.02
17	18	158.86	50.42	43.00	1.72 ± 0.64	3.83 ± 1.05	3.11 ± 1.28	2.08
18	19	164.02	52.72	43.74	1.65 ± 0.33	3.42 ± 1.19	3.71 ± 1.06	2.05

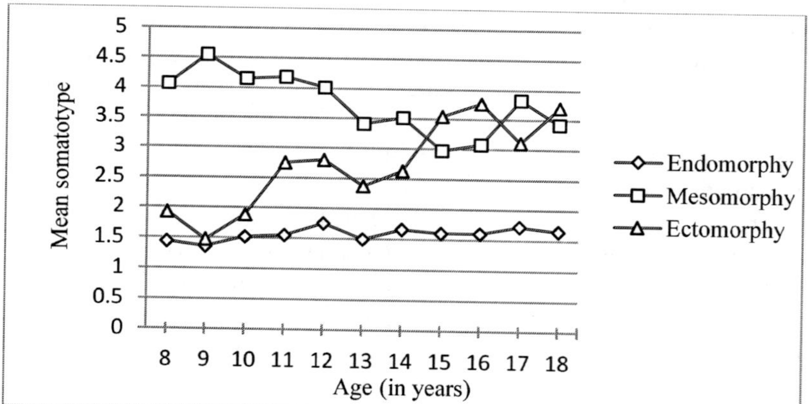

Figure 18.1: Mean somatotype components of Chakhesang Naga boys by age

Table 18.2 gives the differential values using a student's t-test between the preceding and the succeeding age group for the three somatotype components. From the table, it is clear that even though there were minor changes in the endomorphy component ratings, no significant variation was observed between all the age groups compared. While in mesomorphy, significant variation at a 5.0 percent level of probability was found between age groups 8 and 9 years and 12 and 13 years. In the ectomorphy component, significant variation was observed between the age group 14 and 15 years. Between 8 and 18 years of age groups, significant variation was found in mesomorphy and ectomorphy components.

Table 18.2: Values t-test of the three somatotype components of the Chakhesang Naga boys by age

Age group (in years)	*Endomorphy*	*Mesomorphy*	*Ectomorphy*
8 and 9	0.05	2.66*	1.09
9 and 10	0.14	1.56	0.91
10 and 11	0.33	0.10	1.85
11 and 12	1.35	0.62	0.11
12 and 13	1.58	2.39*	1.07
13 and 14	1.32	0.36	0.64
14 and 15	0.29	1.84	2.36*
15 and 16	0.00	0.30	0.56
16 and 17	0.61	1.92	1.52
17and 18	0.42	1.04	1.44
8 and 18	1.90	2.32*	4.84*

* Significant at 5% level of probability

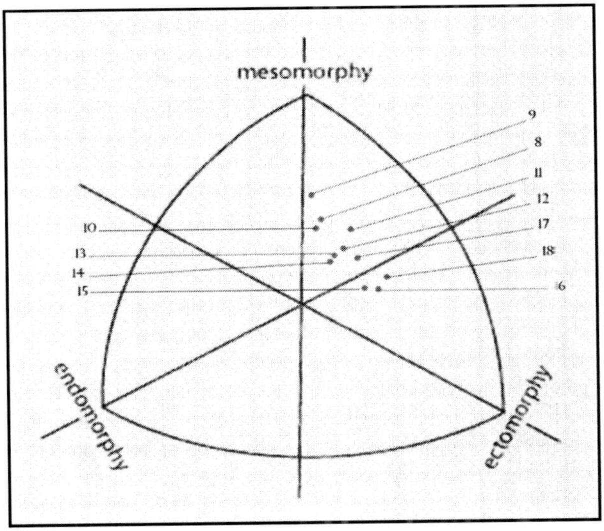

Figure 18.2: Mean somatotype categories of Chakhesang Naga boys at each age group

Somatotypes of the present study were grouped in different somatotype categories following the component dominance method given by Carter and Heath (1990) as shown in table 18.3. Since there was no individual grouped in five categories *viz.,* balanced-endomorph, mesomorph-endomorph, mesomorphic-endomorph, ectomorphic-endomorph, and endomorphic-ectomorph, these categories have been ruled out for inclusion

Table 18.3: Age-wise frequency and percentage of Chakhesang Naga boys in various somatotype categories

Somatotype categories	*Age group (in years)*										
	8	*9*	*10*	*11*	*12*	*13*	*14*	*15*	*16*	*17*	*18*
Balanced mesomorph	12 63.15 %	17 94.44 %	12 66.67 %	7 33.88 %	8 44.45 %	6 33.33 %	8 44.44 %	4 22.22 %	2 11.11 %	6 33.33 %	5 26.32 %
Balanced ectomorph	-	-	-	-	-	-	1 5.55%	7 35.88 %	5 27.77 %	3 16.66 %	5 26.32 %
Mesomorph ectomorph	5 26.32 %	1 5.56%	2 11.11 %	11 61.11 %	6 33.34 %	1 5.56%	7 38.88	5 27.77 %	7 38.88 %	7 38.88 %	6 31.58 %
Endomorph ectomorph	-	-	-	-	-	-	-	-	-	-	1 5.26%
Endomorphic mesomorph	2 10.53 %	-	-	-	-	-	-	-	-	2 11.11 %	-
Ectomorphic mesomorph	-	-	**2** 11.11 %	-	**2** 11.11 %	**3** 16.66 %	-	-	**3** 16.66 %	-	**1** 5.26 %
Mesomorphic ectomorph	-	-	**2** 11.11 %	-	**1** 5.55 %	**6** 33.33 %	-	**1** 5.55 %	**1** 5.55 %	**2** 11.11 %	**1** 5.26 %
Central	-	-	-	-	**1** 5.55 %	**2** 11.11 %	**2** 11.11 %	**1** 5.55 %	-	-	-

in the table. Somatotypes of the present sample of different age groups are distributed into eight different categories. Balanced-mesomorph and mesomorph-ectomorph categories are present in all age groups studied, forming a dominant character among these boys. The highest incidence of occurrence of balanced-mesomorph is seen among age groups 8, 9, 10, 12, 13, and 14 years. Age group, 9 years is seen as the single highest age-wise category with 94.44 percent individuals falling under this category. Balanced-ectomorph is a dominant character in the age group 15 years. The highest incidence of occurrence of mesomorph-endomorph was observed in age groups 11, 16, 17, and 18 years. Age group, 11 years is the single highest age-wise category with 61.11 percent boys falling under this category. On the other hand, mesomorphic-ectomorph showed equal occurrence (33.33% each) with balance-mesomorph in the age group 13 years. Age group 18 years showed the most variation in various component dominance categories, while the age group 9 and 11 years showed the least variation. It is seen that balanced-mesomorph is a dominant pattern in younger boys and as age increases, the ectomorph and its sub-categories tend to become dominant. Such a trend is also observed in figure 18.2.

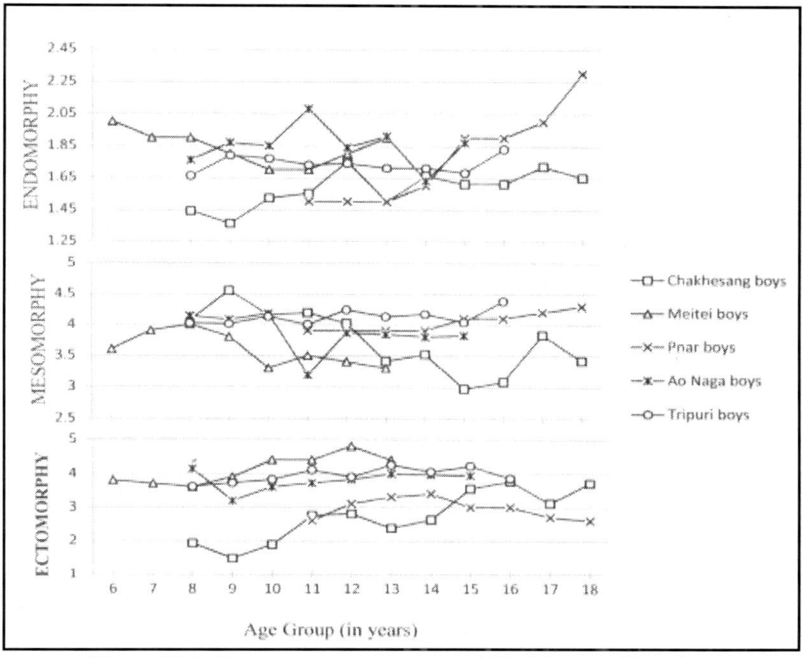

Figure 18.3: Somatotype components of Chakhesang Naga boys and other populations of North-east India.

Considering one-half units of difference in somatotype rating as practically significant (Carter, 1980), the present study has been compared with other reported studies on boys' samples of different populations of North-east India. Similar age group samples for comparison were derived from previously reported studies among Meitei boys of Imphal district, Manipur (Gaur *et al.,* 1999); Pnar boys of Jaintia Hills district, Meghalaya (Dkhar, 2005); Ao Naga boys of Mokokchung district, Nagaland (Longkumer, 2014); Tripuri boys of Tripura (Sarkar and Sil, 2014), and are presented in table 18.3 and figure 18.4. The Chakhesang Naga boys show higher mesomorphy and lesser endomorphy and ectomorphy ratings than the Meitei boys across all age groups compared. Using the criterion mentioned above, no significant variation was observed in the endomorphy component. In mesomorphy, variation was seen in age groups 9 to 12 years. Ectomorphy showed significant variation in all the age groups compared. In comparison with the Pnar boys, the present sample showed generally lesser ratings in the endomorphy and mesomorphy as the age increases, whereas, in ectomorphy, the present study shows a higher rating than the Pnar boys at higher age groups. Age-wise variation in endomorphy (18 years), mesomorphy (15, 16, and 18 years), and ectomorphy (13 to 16 and 18 years) were also observed. Interestingly no variation in the three somatotype components was found in three age groups *viz.,* 11, 12, and 17 years indicating that the present sample is quite similar in the body built to the Pnar population. Between the present sample and Ao Naga boys, no variation was observed in endomorphy while mesomorphy variation was found to be significant only in age groups 11 and 15 years. In contrast, significant variation was found in ectomorphy in all age groups compared to the age group 15 years. Comparison between the present sample and Tripuri boys in respect of endomorphy ratings showed insignificant variation in all the age groups compared. Mesomorphy ratings were significant in higher age groups (13 to 16 years) while in ectomorphy, variation was found absent only in the age group 16 years. On the whole, it can be seen that the present population is quite similar to other populations compared in respect of endomorphy and mesomorphy, but stands quite apart from these populations about ectomorphy ratings showing lesser values. In fact, among these five populations, the present population showed the least average ectomorphy ratings. Since ectomorphy means relative linearity, lower ratings would indicate proper nutrition among the present population. Higher ratings in mesomorphy (relative musculoskeletal robustness) can be attributed to the high consumption of protein-rich food accompanied by active participation in physical activities. Lower ratings in endomorphy (relative fatness) can be attributed to their low intake of fat-rich food products in the form of oil as they prefer boiled vegetables and meat.

Table 18.4: Comparison of somatotype components of Chakhesang Naga boys with other populations of North-east India

Age Group (in years)	*Present study*	*Gaur (1999)*	*Dkhar (2005)*	*Longkumer (2014)*	*Sarkar and Sil (2014)*
	Chakhesang Naga boys	*Meitei boys*	*Pnar boys*	*Ao Naga boys*	*Tripuri boys*
6	-	2.0 - 3.6 - 3.8	-	-	-
7	-	1.9 - 3.9 - 3.7	-	-	-
8	1.44 - 4.07 - 1.92	1.9 - 4.0 - 3.6	-	1.76 - 4.14 - 3.06	1.66 - 4.02 - 3.61
9	1.36 - 4.55 - 1.47	1.8 - 3.8 - 3.9	-	1.87 - 4.08 - 3.19	1.79 - 4.01 - 3.72
10	1.52 - 4.16 - 1.88	1.7 - 3.3 - 4.4	-	1.85 - 4.18 - 3.60	1.77 - 4.13 - 3.82
11	1.55 - 4.19 - 2.75	1.7 - 3.5 - 4.4	1.5 - 3.9 - 2.6	2.08 - 3.19 - 3.72	1.73 - 4.00 - 4.11
12	1.75 - 4.02 - 2.80	1.8 - 3.4 - 4.8	1.5 - 3.9 - 3.1	1.84 - 3.86 - 3.82	1.74 - 4.24 - 3.90
13	1.50 - 3.41 - 2.38	1.9 - 3.3 - 4.4	1.5 - 3.9 - 3.3	1.91 - 3.84 - 4.00	1.71 - 4.13 - 4.26
14	1.66 - 3.52 - 2.63	-	1.6 - 3.9 - 3.4	1.63 - 3.80 - 3.97	1.71 - 4.17 - 4.06
15	1.61 - 2.97 - 3.55	-	1.9 - 4.1 - 3.0	1.87 - 3.82 - 3.94	1.68 - 4.04 - 4.23
16	1.61 - 3.08 - 3.77	-	1.9 - 4.1 - 3.0	-	1.83 - 4.38 - 3.86
17	1.72 - 3.83 - 3.11	-	2.0 - 4.2 - 2.7	-	-
18	1.65 - 3.42 - 3.71	-	2.3 - 4.3 - 2.6	-	-

From figure 18.4, it is clear that the present population is generally ectomorphic-mesomorph and stands very close to the Pnar population in the body built. This proximity between these two populations maybe because of the overall similar age groups taken into consideration for the study. Whereas, Ao Naga boys, Meitei boys, and Tripuri boys fall under the category of mesomorph-ectomorph. These population variations in body physique may be due to certain altitudinal factors, genetic factors, time of sexual maturation, and socio-economic factors. However between very close populations by geography, as in the case of North-east Indian populations, variations may have to do with different dietary patterns in different tribal populations.

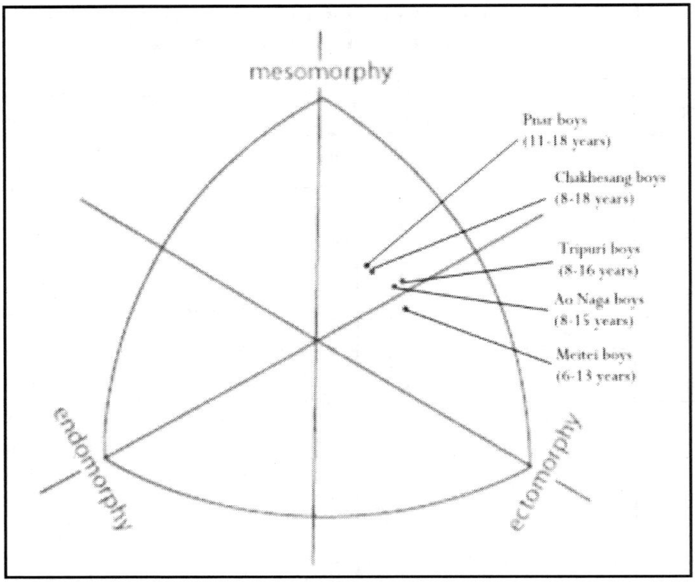

Figure 18.4: Mean somatotype of the Chakhesang Naga boys and other populations of North-east India.

Conclusion

The average somatotype of the Chakhesang Naga boys is 1.58–3.75–2.73 and is generally mesomorphic-ectomorph. The results of the present study show a clear age-related change of mean somatotype components among the boys from age 8 to 18 years. Endomorphy increases from 8 to 18 years but there is no regular trend of this increase was observed. However, in mesomorphy and ectomorphy significant variation was observed. Mesomorphy tends to decrease with age while ectomorphy tends to increase as age progresses. Balanced-mesomorph is a dominant category in younger boys but as age increases, sub-categories of ectomorphy becomes dominant. The present population shows a general trend of similarity with all the populations compared, with the Pnar population being the closest.

References

Bailey, DA, Carter, JEL & Mirwald, RL 1982, 'Somatotypes of Canadian men and women', *Human Biology,* vol.54, no.4, pp.813-828.

Bhasin, MK & Jain, S 2007, 'Biology of the tribal groups of Rajasthan, India: Body mass index as an indicator of nutritional status', *Anthropologist*, vol.9, no.3, pp. 165-175.

Carter, JEL 1980, *The Heath-Carter somatotype method, 3rd edition*. San Diego State University Syllabus Service: San Diego.

Carter, JEL 2002, *The Heath-Carter anthropometric somatotype: Instruction manual* at http://www.somatotype.org/HeathCarterManual.pdf

Carter, JEL & Heath, BH 1990, *Somatotyping: Development and Applications*. Cambridge: Cambridge University Press.

Dkhar, JW 2005, 'Somatotypes among the Pnar boys of Meghalaya'. *Bulletin of Anthropology, Dibrugarh University, Assam,* vol.*33*, pp.85-92.

Gakhar, I & Malik, SL 2002, 'Age changes and sex differences in somatotypes among Jats of Delhi', *Anthropologist* (Special Issue), vol. 1, pp.115-125.

Gaur, R, Singh, SG & Lakhanpal, M 1999, 'Somatotypes of urban Meiteis of Imphal, Manipur', *Anthropologist*, vol.1, no.4, pp. 235-240.

Ghosh, S & Malik, SL 2010, 'Variations of body physique in Santhals: An Indian tribe', *Collegium Antropologicum,* vol.34, no.2, pp. 467-472.

Ghosh, S & Malik, SL 2004, 'A comparative study of age changes in somatotypes of Brahmin and Rajput boys of Sundarnagar, Himachal Pradesh', *Anthropologist,* vol.6, no.1, pp. 19-23.

Handa, N, Karir, BS & Kaul, S 1995, 'Somatotype change during adolescence in Jat Sikh boys of Chandigarh,' *Journal of Indian Anthropological Society*, vol.*30*, pp.85-88.

Heath, BH & Carter, JEL 1967, 'A modified somatotype method', *American Journal of Physical Anthropology,* vol.27, no.1, pp. 57-74.

Heath, BH & Carter, JEL 1971, 'Growth and somatotype patterns of Manus children, territory of Papua and New Guinea: Application of a modified somatotype method to the study of growth pattern', *American Journal of Physical Anthropology*, vol. 35, pp. 49-67.

Kaul, S, Karim, BS & Nand, N 1996, Somatotype variation in India. In: LS Sidhu, SP Singh (Eds.): *Human Biology: Global Developments*. Ludhiana: USG Publishers and Distributors, pp. 61-82.

Kaur, M 2009, 'Age changes in somatotype components of rural and urban Punjabi Brahmin females', *Journal of Human Ecology*, vol.*25, no.*3, pp. 167-173.

Kaur, T & Pathak, RK 2014, 'Somatotype variation in relation to sexual maturation of adolescent girls of Sirmaur District, Himachal Pradesh', *Journal of Life Science*, vol.*6*, pp.41-45.

Longkumer, T 2014, 'Physical activity and somatotypes among Ao Naga boys', *The Anthropologist*, vol.17, *no.*2, pp. 669-675.

Parizkova, J & Carter, JEL 1976, 'Influence of physical activity on somatotypes in boys', *American Journal of Physical Anthropology*, vol.44, pp. 327-340.

Sarkar, SR & Sil, SK 2014, 'Somatotype of non-athlete tribal schoolboys of West Tripura District, Tripura', *Biolife,* vol.2, no.4.

Sarkar, SR & Sil, SK 2014, 'Somatotype of Tripuri tribal adolescence boys (aged 8-16 years) of Tripura, North-east India', *Proceedings International Ergonomics Conference*, pp. 727-731.

Sarkar, SR & Sil, SK 2015, 'Body physique in Bengali adolescent boys from west Tripura district, Tripura, India', *International Journal of Bioassays*, vol.4, no.3, pp. 3684-3688.

Singh, LD 2011, 'Somatotypes of the affluent and non-affluent Meitei boys of Manipur, India', *Anthropologist*, vol.13, *no.*1, pp. 9-16.

Singh, SP & Sidhu, LS 1980, 'Changes in somatotypes during 4 to 20 years in Gaddi Rajput boys,' *Zeitschrift für Morphologie und Anthropologie*, vol.*71, no.*3, pp.285-293.

Singh, SP, Singh, P, Malhotra, P & Sidhu, LS 2007, 'Somatotypes of high altitude Spitian boys', *Journal of Human Ecology*, vol.*22, no.*2, pp. 129-133.

Thapliyal, N & Kumar, A. 2019, 'Comparison of somatotype of physically active Punjabi men of different age group', *Journal of Exercise Science and Physiotherapy*. Vol.15, no.1, pp. 23-28.

Weiner, JS & Lourie, JA 1981. *Practical Human Biology*. New York: Academic Press.

Chapter - 19

Determinants of Fertility in North-eastern India: A Case of the Lepchas

A.K. Gyanendra Singh Wahengbam and Benrithung Murry

Introduction

Fertility is a generic concept that encompasses a complex time-dependent process covering events ranging from exposure to coitus and childbearing to the formation and dissolutions of unions. It is a complex multivariate system representing a set of different reproduction strategies (McNicoll, 1980). Though reproduction is a biological phenomenon, levels of fertility can be influenced by socially control norms and practices. Fertility is one of the main components of population growth. It has an important role to play in the age structure of the population. Young or old age structure depends on the rise or decline in corresponding birth rates. Therefore, a population stabilizes itself by bringing down the number of births. There is a need to understand the levels of fertility patterns in society and the related factors influencing it.

Factors affecting fertility

Human fertility is influenced by biological, sociological, and economic factors. The impact of social and economic change on reproductive behavior is mediated through the cultural setting. Age at menarche, age at marriage, age at conception, birth order or order of pregnancy, etc. are biological factors that can affect fertility to a large extent. Education, per capita annual income (PCAI), occupation, and birth control fall under socio-economic factors affecting fertility. Generally, fertility among tribal is high and varies from one tribe to another. According to Majumdar (1947), economic conditions are largely responsible for these variations. During his field studies, he collected information on children ever born and children surviving among six tribes. His estimates suggest high fertility among tribals, for instance, Ho (6.2), Oraon (6.0), Kuki (6.5), Tharu (6.6), and Saora (5.7). Shreds of evidence of high fertility among tribal populations have also been noticed in another study in Bihar. Verma (1977) estimated that the average number of children born to ever-married women aged 15-49 years is 6.69 among Santhals and 6.33 among Birhor tribes. A good number of past findings show that the fertility level of the tribal communities could not be reduced to the national level. It is associated with factors like low socio-economic status, disadvantages of communication and infrastructure, etc. (Bhagat and Chattopadhyay, 2004; Nanda, 2005; Saha and Verma, 2006).

Menarche is the onset of the reproductive period in females. The first menstruation is taken into account like the age of puberty or start of reproductive maturity and from this stage, females are biologically capable to conceive (Field and Ambrus, 2008). Mckibben and Poston's (2003) study on Chinese women found a positive association between age at menarche and children ever born, i.e., the earlier a woman's age at menarche, the greater her number of children ever born. A similar result was reported among African women by Presser (1978).

Marriage is an event that marks the beginning of the period of potential childbearing but fertility is affected by a range of biosocial factors. While age at marriage may be considered to be an important variable affecting fertility behavior, educational and literacy level are widely argued as having an inverse relationship with fertility (Caldwell, 1979; Reddy and Caldwell, 1985 and many more). Education also affects the health practices of individuals. There is an inverse association between income per adult and fertility reported from many studies. Reports from different scholars suggest that fertility is lower among better-educated women and is often higher among women those families who own more land and assets. Fertility is often higher in poorer families within a society, and across countries, those with higher average fertility tend to have a lower average income.

The North-east India experience

The total fertility rate (TFR) for the country had declined to 2.2 in 2017 after it remained at 2.3 for four years from 2013 to 2016. The fertility trend has declined throughout the North-eastern states which conforms to the all-India fertility trend as evident from the Sample Registration System (SRS, 2017) of India. However, there is still a marked difference between urban and rural fertility in the North-east region (NFHS-4). This differential fertility trend is largely attributable to certain biological and socio-cultural factors. The aspect of fertility occupies a central position in the study of any population because it is responsible for the biological replacement and maintenance of human society. Fertility is not only a major portion of all demographic behavior but also a fundamental element in social structure and human conditions. In general, fertility is the prime determinant of age structure, family composition, and growth rate.

The study on the demographic behavior of the Indian tribes, particularly of the tribal communities of North-eastern India is limited. A few demographic studies on the Indian tribes were already documented earlier (Mutharayappa, 1994; Basu and Kshatriya, 1997; Kapoor and Kshatriya, 2000; Maharatna, 2000; Pandey and Tiwary, 2000; Biswas and Kapoor, 2003; Nanda, 2005). A review of anthropological literature that examined fertility differentials among different ethnic groups of Sikkim professing Buddhism and Hinduism has given insight that human fertility is influenced by a host of biological, socio-economic, and socio-cultural factors (Bhasin and Bhasin, 2000). The impact of social and economic change in reproductive behavior is mediated through the cultural setting. Indeed, in most ethnic groups there was some recognition of both advantages and disadvantages of smaller family size, although most agreed that under the present circumstance the advantages were prominent. This study is an attempt to comprehensively document the demographic characteristics and the transition of the Lepcha community.

Materials and methods

A preliminary interview schedule was drafted and administered and was subsequently modified after pretesting so that an accurate response could be obtained. The first part of the interview schedule is related to information on proper identification and household composition. The second part of the interview schedule probes into the reproductive profile of the women, immunization and child care, morbidity and mortality, women's health, etc. The third part probes the status of women in the family and society along with attitude towards female child and daughter's education. Finally, the last part deals with the collection of the data relating to sanitation, educational institutions, health care centers to assess the overall development of the area. The present study was conducted in Dzongu, a special reserved area of Lepchas in the North Sikkim district. The unit selected for the demographic study was the ever-married women in the reproductive age group of 15-49 years that comprised of 457 women. A purposive sampling method was used and collected data from 14 villages of Dzongu for the present study.

Results

We considered several factors influencing the fertility of women in the Lepcha population which comprises biological factors and socio-economic factors. Some fertility rates of the Lepchas in comparison to the populations of other North-eastern states and India as well are highlighted in table 19.1. The crude birth rate (CBR) of the Lepchas is 22.2 which is higher than the average national value and that of other North-eastern states excepting Meghalaya. The total fertility rate (TFR) of 3.2 among Lepchas is very high as compared to all others. This may be, however, because of the limited sample of Lepchas in comparison to larger SRS samples for states and India. Among all the North-eastern states Meghalaya has the highest crude birth rate and total fertility rate whereas Tripura has the least crude birth rate and total fertility rate. Assam and Sikkim are the only states which come close to the national level in the total fertility rate.

Table 19.1: Distribution of crude birth rate and total fertility rate

India and States	*Crude birth rate (CBR)*	*Total fertility rate (TFR)*	*Reference(s)*
Assam	21.2	2.3	SRS 2017
Arunachal Pradesh	18.3	-	SRS 2017
Manipur	14.6	1.5	SRS 2017, 2013
Meghalaya	22.8	2.7	SRS 2017, 2013
Mizoram	15.0	1.5	SRS 2017, 2013
Nagaland	13.5	1.6	SRS 2017, 2013
Tripura	13.0	1.3	SRS 2017, 2013
Sikkim	16.6	2.0	SRS 2017, 2013
India	20.2	2.2	SRS 2017
Present Study	22.2	3.2	Present study

Table 19.2: Distribution of fecundity and infecundity

Age-cohort	*No. of ever-married women*	*No. of Fecund women*	*%*	*No. of Infecund women*	*%*	*Newly married*	*%*
15-19	11	5	1.15	-	-	6	1.38
20-24	76	68	15.60	1	0.23	7	1.61
25-29	103	98	22.48	-	-	5	1.09
30-34	92	91	20.87	1	0.23	-	-
35-39	89	89	20.41	-	-	-	-
40-44	49	49	11.24	-	-	-	-
45-49	37	36	8.26	1	0.23	-	-
Total	457	436	95.40	3	0.69	18	4.13

The distribution of fecund and infecund women is shown in table 19.2. Out of the total 457 women, 436 women constituting 95.40 percent are fecund and a negligible 0.69 percent of women are in-fecund, while 4.13 percent of women are newly married.

Table 19.3: Factors affecting conceptions and live-birth among the ever-married Lepcha women.

Characteristics	*No. of women*	*Conception*		*Live-Birth*		*p-value*
		Total Conception	*Mean Conception*	*Total Live-Birth*	*Mean Live-Birth*	
Literacy status						
Literate	379	888	3.1	867	[illegible].02	0.968
Non-literate	78	242	1.91	238	[illegible].03	
Occupational Status						
Farmer	72	185	3.03	181	3.01	0.999
Cultivator	326	790	2.42	772	2.37	
Salaried	59	155	2.64	152	2.61	
Economic Status						
Below poverty line	444	1093	3.04	1070	3	0.889
Above poverty line	13	37	2.38	35	2.29	
Antenatal Care (ANC)						
Complete	94	141	1.8	137	1.78	0.978
Incomplete	342	989	2.57	969	2.51	
Place of delivery						
Hospital	251	491	2.03	489	2.02	0.702
Home	185	639	3.2	616	3.19	
Delivery Conducted by						
Medical personnel (doc/nurse)	293	651	2.93	639	2.89	0.925
Untrained attendant (dai)	143	479	3.25	466	3.21	
Lactational amenorrhea						0.57
6 months	236	712	2.19	709	2.18	
Less than 6 months	200	418	3.47	396	3.46	
Postnatal Care (PNC)						
Complete	59	16	2.72	16	2.65	0.943
Incomplete	377	1124	3.69	1089	3.61	
Birth Control Measures						
User	381	976	2.48	976	2.4	0.164
Non-user	76	154	2.2	129	2.1	

Literate women are found to have higher mean conception than non-literate women but both have similar mean live birth (Table 19.3). Women who practice farming and who are below the poverty line have higher fertility among Lepchas. Higher fertility is also seen among women who received incomplete antenatal care and postnatal care. Women who have institutional delivery have lower fertility than those women whose delivery was conducted by untrained attendants. Those women who have undergone lactational amenorrhea for less than 6 months and who use birth control measures have also higher fertility. Thus, it is seen that literacy level, occupation, economic conditions, birth control measures, and access to medical facilities do not influence the fertility level of Lepchas. This is also evident from the p-value where no parameter considered is found to be statistically significant. This suggests that some socio-cultural beliefs and practices are operating in the population which has a far greater impact on fertility status in this tribal group.

Table 19.4: Multivariate stepwise regression analysis of the factors affecting fertility

Sl.No.	*Parameters*	*R Square*	*p-value*	*SE*
1.	Age at menarche	52.2	0.011	0.038
2.	Age at marriage	51.8	0.000	0.007
3.	Age at first conception	50.4	0.000	0.005
4.	Educational status	52.2	0.479	0.069
5.	Occupational status	52.3	0.630	0.017
6.	Economic status	52.3	0.542	0.000
7.	Birth control measure	52.2	0.252	0.351

Multivariate stepwise regression analysis was done by choosing a few bio-social parameters to understand the overall influence of different determinants of fertility and their level of significance in the contribution of fertility differentials as shown in table 19.4. All the chosen parameters have almost an equal share of influence on fertility ranging from 50.4 to 52.3 percent. The age at marriage and age at first conception, however, have a statistically significant influence on the overall factors affecting fertility.

Discussion

The most significant finding from this study is that the socio-cultural factors are found to have more influence on the fertility among the Lepchas. This is also true for some tribal populations in North-east India as reported from many earlier studies. A comparison among the literate and non-literate women found that literate women have higher mean conception and live-birth against those women who are non-literate. This finding is, however, deviating from the previous studies done by Basu (2002) and Adhikari (2010) where they found that literacy has a direct influence on the fertility of women since educated women are more aware of birth control measures as well as they have autonomy on reproductive decision-making which helps to lower their fertility. Considering the occupational status, those women who practice agriculture have higher mean conception and live-birth as against the salaried women whose mean conception and live-birth are lower. Similarly, women who are below the poverty line have higher fertility as compared to women above the poverty line. This concord with the earlier study of Al-Kandari (2007) and another study made by Hezarjaribi and Abbaspoor (2010). Women who have complete antenatal care have lesser fertility as compared to those who have incomplete antenatal care. In the present study, the number of women receiving complete antenatal care is very less and this is influenced by the inclination to consult traditional health practitioners in most of the tribal populations as reported by Negi *et al.*, (2010). Those women who received complete postnatal care is also very less and those who received incomplete care have more fertility.

Considerable policy attention was given to maternal health in the Millennium Development Goal (MDG), including the introduction of the *Janani Suraksha Yojana* (JSY) in 2005, a national conditional cash transfer scheme to incentivize institutional deliveries. Health facilities in rural areas were also strengthened through the National Rural Health Mission (NRHM). Also, a community health volunteering programme called Accredited Social Health Activists (ASHAs) was introduced to support and encourage pregnant women to deliver in public health facilities. *Janani Shishu Suraksha Karyakram* (JSSK, or mother and child safety programme) was launched in 2011 to address high out of pocket expenses to remove the barrier to institutional delivery. Consequently, the utilization of maternal health care, as well as institutional deliveries, increased dramatically. This transition is much visible in the younger women in tribal populations as the majority of them get to avail of such facilities and these younger age groups are found to have lower fertility as compared to the women in the older age group.

Interestingly, fertility among the Lepchas is higher among women who use birth control measures. It appears that there is no effect of the use of birth control measures on the fertility of women and many couples opt for birth control measures after they have had the desired family size.

This observation is similar to the study made by Chandiok *et al.,* (2016). The present studied population has extensive use of contraception. However, few previous studies (Kapoor and Kshatriya, 2000; Borah, 2005; Borah, 2006) had reported poor use of contraceptives among the Indian tribes.

Conclusion

Most of the North-eastern states have witnessed a decreasing trend of total fertility rate which is much below the replacement fertility of the country except in the states of Assam and Meghalaya, where the total fertility rate is much higher. Literacy and increased use of contraception are believed to have played a role in the decline of the fertility rate in the region. But, the fertility dynamics among the Lepchas of Dzongu in North Sikkim are deviating from other North-eastern states as there is a very high total fertility rate even though the majority of the women are literate and availing of contraceptive measures. However, early age at marriage which is common to the tribal populations coupled with adherence to cultural practices and beliefs are found to have impacted the fertility behavior of this tribal group in North-east India.

References

Adhikari, R 2010, 'Demographic, socio-economic, and cultural factors affecting fertility differentials in Nepal', *BMC Pregnancy and Childbirth*, vol.10, no.1, pp.19.

Al-Kandari, Y 2007, 'Fertility and its relation with socio-cultural factors in Kuwaiti society', *Eastern Mediterranean Health Journal*, vol.13, no.6, pp. 1364-1371.

Basu, A 1998. *Cultural material and reproductive behaviour: A new approaches to anthropological demography.* Oxford; Oxford University Press.

Basu, AM 2002, 'Why does education lead to lower fertility? A critical review of some of the possibilities', *World Development*, vol.30, no.10, pp. 1779-1790.

Basu, S & Kshatriya, GK 1997, 'Fertility and mortality trends in the Kharia tribals of Odisha,' *Social Change*, vol. 27, no. 1, pp. 114–128.

Bhasin V & Bhasin MK. 2000. A cross-cultural fertility differential in Sikkim. *Journal of Human Ecology*, vol. 11, no 6, pp 429-444.

Biswas, RK & Kapoor, AK 2003, 'Fertility profile of a primitive tribe, Madhya Pradesh,' *Anthropologist*, vol. 5, no. 3, pp. 161–167.

Biswas, RK & Kapoor, AK 2003, 'Age at menarche and menopause among Saharia women- A primitive tribe of Madhya Pradesh', *Anthropologist*, vol.6, pp.247-252.

Census of India 2011, Provisional Population Totals- India- Data Sheet.

Chandiok, K, Mondal, PR, Mahajan, C & Saraswathy, KN 2016, 'Biological and social determinants of fertility behaviour among the Jat women of Haryana state, India', *Journal of Anthropology*, 5463168. Available at: http://dx.doi.org/10.1155/2016/5463 168.

Caldwell JC. 1979. Education as a factor in mortality decline, an examination of Nigerian data. *Population Studies*, vol. 33, no.3, pp 395-413.

Field, E and Abrus, A 2008. Early marriage, age of menarche, and female schooling attainment in Bangladesh. *J.Polit. Econ*, 116 (5):881-930. 10./086/593333

Hezarjarib, J & Abbaspoor, A. 2010, 'Influence social-economic on women fertility rate', *Social Research*, vol. 3, pp.178-184.

Kapoor, AK & Kshatriya, GK 2000, 'Fertility and mortality differentials among selected tribal population groups of north-western and eastern India,' *Journal of Biosocial Science*, vol. 32, no. 2, pp. 253–264.

Maharatna, A. 2000, 'Tribal fertility in India: socio-cultural influences on demographic behaviour,' *Economic and Political Weekly*, vol. 35, no. 34, pp. 3037-3047.

Majumdar, DN. 1947. *The Matrix of Indian Culture*. Universal Publishers Ltd, Lucknow.

Mckibben, SL & Poston, DL 2003, 'The influence of age at menarche on the fertility of Chinese women', *Journal of Social Biology*, Autumn-Winter; vol.50, no. (3-4), pp.222-37, the University of Texas of the Permian Basin, USA.

McNicoll, G 1980, 'Institutional determinants of fertility change', *Population and Development Review*, vol.6, no.3, pp. 441-462.

Nanda, S 2005, 'Cultural determinants of human fertility: A study of tribal population in Odisha', *Anthropologist*, vol.7, no.3, pp. 221-227.

National Family Health Survey (NFHS-4). India. Sikkim State Report. 2015-2016

Negi, NS, Sekher, TV& Ganguly, S 2010, 'Antenatal care among tribals: A study of Chhattisgarh and Jharkhand', *Studies of Tribes and Tribals,* vol. 8, no.2, pp.77-86.

Pandey, GD & Tiwary, RS 2000, 'Some aspects of population growth of the primitive tribes of Madhya Pradesh,' *The Journal of Family Welfare*, vol. 46, no. 2, pp. 66–72.

Saha, K & Verma, B 2006, 'High fertility among scheduled tribes of Madhya Pradesh', *Indian Journal of Medical Research*, vol.123, pp.89-90.

Chapter - 20

A Multidisciplinary Approach to COVID-19 with Special Reference to North-east India: Genetic Factors and Beyond

Kusum Ghosh, Abhishikta Ghosh Roy, Diptendu Chatterjee, Arup Ratan Bandyopadhyay

COVID-19 is a novel infectious, viral disease, caused by SARS-CoV-2 which belongs to the family Coronaviride (Gorbalenya *et al.,* 2020), and WHO declared as a pandemic. This disease endangers disproportionately the males (Jin *et al.,* 2020) and the elderly especially those with pre-existing co-morbidities (Yang *et al.,* 2020). From the large family of Coronaviruses, three have been known to cause severe pneumonia such as Severe Acute Respiratory Syndrome (SARS; Drosten *et al.,* 2003) and Middle East Respiratory Syndrome (MERS; Zaki *et al.,* 2012) and recently recognized Severe Acute Respiratory Syndrome- 2 (SARS-CoV-2; Huang *et al.,* 2020). The transmission occurs from one person to another through the respiratory droplets, produced when an infected person coughs or sneezes in the close presence of other people (Huang *et al.,* 2020; Peeri *et al.,* 2020). Yet, no effective drugs or vaccines are clinically approved to prevent and cure this viral infectious disease. Hence, it is pertinent and imperative to unravel strategies (Chen *et al.,* 2020) for prevention. SARC-CoV-2 infection possesses varying degree of susceptibility and lethality both between countries and also within counties across different regions (Dhillon *et al.,* 2020; Bandyopadhyay *et al.,* 2020). Although India is at 3rd place (as of 21st July 2020) based on confirmed cases but Case Fatality Rate (CFR) of India is still very low in comparison to worldwide prevalence, not only that observation of different parts of India revealed several confirmed cases and even the CFR in the North-east is less than the other parts of India (COVID-19 tracker, 2020), pointing to some innate, natural defense against SARS-CoV-2. Reasons underlying these differences in CFR both between and within-population remain uncertain and are likely to be multifactorial and driven by a combination of Genetic and Epigenetic factors (Public Health England, 2020).

Genetic factors and Covid-19 variations

Researches demonstrated that SARS-CoV-2 enters the human body through Angiotensin-Converting Enzyme-2 (ACE2) receptor, present in the lung epithelial cell. The genetic predisposition for an increased risk of SARS-CoV-2 infection, related to ACE2 expression. Studies reported higher expression of ACE2 is responsible for the disease prevalence and severity (Zhao *et al.,* 2020) which is also known to require for maintaining the Renin-Angiotensin-Aldosterone-System (RAAS), act as a homeostatic regulator of vascular function (Tikellis *et al.,* 2012). Although SARS-CoV-2 infection decreases the level of ACE2 (Chen *et al.,* 2020) and develops Cytokine storm (Song *et al.,* 2020) as a means of autoimmune response along with Acute Respiratory Distress Syndrome (ARDS) (Gattinoni *et al.,* 2020) but who possesses a high level of ACE2 expression, counteracted this by their higher basal ACE2 level thus, have protection from this infectious disease even though they have higher susceptibility (Chen *et al.,* 2020). This means severity arises among those having reduced expression of ACE2 (Devaux *et al.,* 2020) which is highly dependent on the Angiotensin Converting Enzyme (ACE) polymorphism (Pinheiro *et al.,* 2019). ACE2 is a distinct type but a close homologue of ACE. ACE converts the hormone Angiotensin I to the active vasoconstrictor Angiotensin II and it also degrades bradykinin, acting as a vasodilator. Therefore, ACE indirectly increases blood pressure by causing blood vessels to constrict. But ACE2 helps in the degradation of Angiotensin II, resulting in the formation of vasodilator Angiotensin (1-7) which opposes the action of Angiotensin II and maintains the vascular function (Tikellis *et al.,* 2012). But the expression of this ACE2 is decreased among the individuals who possess the "D" allele (ACE1 (I/D) polymorphism), whereas, approximately doubled the ACE concentration (Pinheiro *et al.,* 2019) which in conjugation results in several comorbidities including, Hypertension, Diabetes, Cardiovascular disease (Gard, 2010) thus, the severity of COVID-19 infection (Yang *et al.,* 2020). The ACE I/D polymorphism discerned important geographical variation (Saab *et al.,* 2007). Although the outbreak of SARS-CoV-2 emerged in Asian countries but after that European countries and the USA experienced a much higher incidence of morbidity and mortality rate (COVID-19 tracker, 2020). Studies reported (Kenyon *et al.,* 2020; El Ezzi *et al.,* 2020) frequency of the D-allele is higher in Europe, Southern Europe, and the USA, compared to Asia and particularly India, hoist intuitive phenomenon behind the higher frequency of morbidity and mortality among the Europeans than in the Asians. Nevertheless, studies from different regions of the Indian population on ACE (I/D) polymorphism revealed higher I allele frequency in the North-eastern region (Ghosh *et al.,* 2018), which can be a reason for the lower prevalence of SARS-CoV-2 in the North-east Indian population (COVID-19 Tracker, 2020).

In search of epidemiological concern and population variation in the susceptibility of COVID-19, a recent study (Zhao *et al.,* 2020) reported ABO blood group displayed different association risk in driving disease severity among COVID-19 patients, where individuals with "A" blood group have a higher chance of being affected, while the rate of infection with blood type "O" is minimal. The locus that encodes blood type and a multi-gene cluster on Chromosome 3, encodes for a transporter protein known to interact with ACE2 and potentially have relevant functions to severe Covid-19 and also encodes several Chemokine receptors, plays role in Immune response. Researchers noted that genetic variants of these proteins associated with increased levels of Cytokine: Interleukin-6 (IL-6) plays a role in inflammation and may have implications for COVID-19 severity (Ellinghaus *et al.,* 2020). Furthermore, ABO blood group also varies geographically where the studies suggest the higher frequency of blood type "O" in the North-eastern part of India than the other regions (Banerjee, 1989; Bandyopadhyay, 1993; Agarwal *et al.,* 2014), might be a reason of lower CFR due to COVID-19.

Table 20.1: Region-wise distribution of 'D' allele prevalence and associated Case Fatality Ratio*

Region (INDIA)	*Number of Confirmed Cases*	*Number of Recovered Cases*	*Case Fatality Rate (CFR)*	*Recovery Rate*	*ACE I/D*	*Prevalence of Blood Group "O"*
West	398789	233006	3.70%	58.43%	D	36.75%
Central	28908	19628	2.65%	67.90%	D	43.24%
North	219739	168379	2.52%	76.63%	D	29.43%
East	97156	59850	1.58%	61.60%	I	37.55%
South	376970	221637	1.44%	58.79%	I	40.00%
North-East	33282	21355	0.21%	64.16%	I	60.00%

*COVID 19 Tracker accessed on July 21, 2020.

Studies tried to identify the culprits and mechanisms for the observed differentials in COVID-19 fatality rates in a different age, sex, and ethnic groups from a biological specifically genetics point of view (Bandyopadhyay *et al.,* 2020). But some factors are also there which is beyond genetics i.e. epigenetic factors like, Demographic structure, Population Density, Co-residency pattern, etc. associated with the susceptibility of COVID-19 (Esteve *et al.,* 2020; Singh *et al.,* 2020; Rocklov *et al.,* 2020; Dudel *et al.,* 2020).

Demographic structure

The Demographic structure of a population is a key determinant of a pandemic's progression (Hope-Simpson, 1970). Demographic science can provide new insights into how the pandemic may unfold, its intensity (Dowd *et al.,* 2020), and also how the changes in the demographic structure of a population affect disease transmission (John, 1990; Manfredi *et al.,* 2004). The demographic composition of a population, specifically, population age structure may explain the remarkable variation in fatalities both across and within-population (Dowd *et al.,* 2020; Verity *et al.,* 2020). Over time the Demographic structure of populations is changing with an increase in life expectancy and declining fertility (Geard *et al.,* 2015; Livi-Bacci, 1997). This on the other hand leading to population ageing, characterized by an increasing mean age of a population i.e. elderly persons are increasingly represented within a country's overall age structure (Geard *et al.,* 2015; Shrestha, 2000). Population ageing is a human success story that reduces early deaths and increases life expectancy by the advancement of economic, social, medical, and public health condition (United Nation, 2019) but evidence shows that the risk of severe diseases and mortality increases sharply with age because of lower immunity in the older people (WHO, 2020; Verity *et al.,* 2020). Several pieces of research on COVID-19 also revealed that the mortality rate is higher among the older people across the world (Huang *et al.,* 2020; Lithander *et al.,* 2020; Nanda *et al.,* 2020) thus, the regions with higher mean age are at increased risk of severity of COVID-19 infection.

Population aging is currently more pronounced in developed countries like Europe, North America, and Central America as well as in less developing countries like Eastern and South-eastern Asia (Garden *et al.,* 2015). The regions like America, Europe, China, with narrower base i.e. Lower percentage of younger people, typical of aging populations at or near sub-replacement fertility exacerbate their vulnerability to COVID-19 related deaths (Esteve *et al.,* 2020; Singh *et al.,* 2020) than the regions (Africa and India) with Taj Mahal dome shape population structure typical of younger population (Dowd *et al.,* 2020; Singh *et al.,* 2020; Dhillon *et al.,* 2020). In India, special mention can be made of North-east India

(COVID-19 tracker, 2020) where the percentage of the older population (above the age of 60 years) is only 5.3 percent (Census, 2006), which might be the reason for lower death rate due to COVID-19.

From this, it can be concluded that countries with older populations must take aggressive protective measures. Mapping of age-related spatial clustering both within and across countries can improve hospital and critical care forecasts (Verhagen *et al.*, 2020). Demographically informed projections will help the Government to better predict the COVID-19 burden and to implicate necessary measures towards the design of more effective strategies for disease control.

Table 20.2: Distribution of mean age of different worldwide regions

Region	*Mean Age*	*CFR*
Europe	42.6	6.90%
United States of America	38.2	3.55%
China	38.4	5.41%
Africa	19.7	2.14%
India	26.8	2.02%

Table 20.3: Distribution of aged people in North-east India

Region	*Below 50 (Age) population (%)*	*Above 50 (Age) population (%)*
North-east India	94.7%	5.3%

Population density

Population density is another key factor that is responsible for the severe outcome in the case of any pandemic. A higher density of population makes the great cities more productive and innovative and also unknowingly makes them vulnerable to infectious diseases (Li *et al.*, 2018). Across the world, COVID-19 has taken root and hit hard the places with higher population density, for example, the industrial centers like Wuhan, Detroit, and Northern Italy (Huang *et al.*, 2020; Kenyon *et al.*, 2020), which are connected through supply chains and also New York and London (Yang *et al.*, 2020), marked with a large flow of visitors and tourists, diverse global populations and dense residential areas. This scenario is consistent with that of India. No places seem to be immune till now except North-east India where the number of cases is still relatively lower than the other counterparts (COVID-19 Tracker, 2020) might be due to the lowest population density per square kilometer (Census, 2011). But some states were found to be more vulnerable to the virus's impacts for example, the most densely populated single place in the world, the Dharavi slum in Mumbai, India. Data revealed that in India only Mumbai's COVID-19 tally is now more than that of China (COVID-19 Tracker). Regions with dense populations are particularly marked with severe infectivity because keeping the distance of more than one-meter between people while coughing and sneezing, becomes more difficult in populations with higher densities (Rocklov *et al.,* 2020; Li *et al.,* 2018; WHO, 2020).

Therefore, to reduce community spread it will be necessary to strategically reduce contact-rates. Controlling contact rates is key to outbreak control, and such a strategy depends on population densities.

Table 20.4: Case Fatality Rate (CFR) according to population density

Region	*Population Density per sq. km.*	*Number of Confirmed Cases*	*Number of Recovered Cases*	*Case Fatality Rate (CFR)*	*Recovery Rate*
West	344	398789	233006	3.70%	58.43%
Central	417	28908	19628	2.65%	67.90%
North	267	219739	168379	2.52%	76.63%
East	625	97156	59850	1.58%	61.60%
South	397	376970	221637	1.44%	58.79%
Northeast	176	33282	21355	0.21%	64.16%

Co-residency

Besides the population demographic structure and density, the structure of co-residency i.e. how big are households and how old are their members, recognized as another key factor that determines the vulnerability of the countries to COVID-19 outbreak (Dowd *et al.,* 2020) due to the duration and intensity of contacts occurring within the household members (Hope-Simpson, 1970). This in turn will help the government in making policies that can shape the number of infections and deaths after an outbreak.

In countries like America, Europe, and China where the elderly form a large part of the population and live with their generational peers, avoiding the primary infection of the elderly people will considerably reduce both the direct and indirect deaths by COVID-19 among them. But countries, where the elderly population forms a small part of the population but live together with young people die because of within household infection, which can outnumber the direct deaths (Esteve *et al.,* 2020). India is characterized by the prevalence of three-generation in a single household. This signifies the frequent contact between children, young and grand-parents which increases the possibility of substantial transmission of the virus from one generation to another. Such contacts are less in China and negligible in Europe and America (Singh *et al.,* 2020). Once community transmission is established, countries with high intergenerational contacts may see faster transmissions to high-fatality, as seen in Italy and Spain, leading to higher average CFR (Dudel *et al.,* 2020).

Therefore, while imposing measures for a country like India, a strategy for prevention vis-à-vis clinical management might be that avoiding primary infections of the older person in these countries will be less effective in reducing the number of deaths because the elderly might still get infected by younger household members.

Based on the available data, in conclusion, it is fairly possible to envisage the occurrence and development of SARS-CoV-2 depend on the interaction between the virus and the host combinations and variability. Viral factors include virus type, mutation, viral load, viral titer, and viability of the virus in vitro. On the other hand, the individual's characteristics factor includes genetic factors in terms of genetic polymorphisms, age, gender, demographic status, a physical status which is variable in different ethnic groups in general and specifically in India. These factors all contribute to whether an individual is infected with the virus, the duration and severity of the disease, and the re-infection.

References

Agrawal, A, Tiwari, AK, Mehta, N, Bhattacharya, P, Ravi, Wankhede R, Sunita, S, Tulsiani, S & Kamath, SS 2014, 'ABO and Rh (D) group distribution and gene frequency; the first multicentric study in India', *Asian Journal of Transfusion Science,* 2014 vol.8, no.2, pp.121–125.

Bandyopadhyay, AR, Chatterjee, D, Ghosh, K & Sarkar, P 2020, 'COVID 19: An epidemiological and host genetics appraisal,' *Asian Journal of Medical Sciences,* vol.11, no.3, pp.71-76.

Chen, J, Jiang, Q, Xia, X, Liu, K, Yu Z, Tao W, *et al.,* 2020, 'Individual variation of the SARS-CoV2 receptor ACE2 gene expression and regulation,' *Preprints,* pp. 1-15 (https://doi.org/10.1111/acel.13168).

Chen, N, Zhou, M, Dong, X, *et al.,* 2020, 'Epidemiological and clinical characteristics of 99 cases of 2019 novel coronavirus pneumonia in Wuhan, China: a descriptive study', *Lancet,* vol.395, pp.507-513.

COVID Tracker Official Updates Coronavirus - COVID-19 in India - mygov.iný (Accessed July 21, 2020)

Devaux, CA, Rolain, J, Raoult, D 2020, 'ACE2 receptor polymorphism: susceptibility to SARS-CoV-2, hypertension, multi-organ failure, and COVID-19 disease outcome', *Journal of Microbiology, Immunology, and Infection,* vol.53, pp. 425-435.

Dhillon, P, Kundu, S, Shekhar, C, Ram, U, Dwivedi, LK, Yadav, S, Unisa, S 2020, 'Case fatality ratio and recovery rate of COVID-19: scenario of most affected countries and the Indian States,' *Preprint*: https://www.researchgate.net/publication/340950473. (27 April 2020).

Dowd, JB, Andriano, L, Brazel, DM, *et al.,* 2020, Demographic science aids in understanding the spread and fatality rates of COVID-19. *Proceedings of National Academy of Science (PNAS)* vol.117, no.18, pp.9696-9698.

Drosten, C, Gunther, S, Preiser, W 2003, 'Identification of a novel coronavirus in patients with severe acute respiratory syndrome', *New England Journal of Medicine,* vol.348, pp.1967-1976.

Dudel, C, Riffe, T, Myrskyla, M, *et al.,* 2020, 'Monitoring trends and differences in COVID-19 case fatality rates using decomposition methods: Contributions of age structure and age-specific fatality', *MedRxiv* (doi: https://doi.org/10.1101/2020.03.31.20048397).

El Ezzi, AA, Clawson, JM, El-Saidi, MA, *et al.,* 2020, 'Association of Angiotensin I Converting Enzyme insertion/287 bp Deletion polymorphisms and proliferative prostatic diseases among Lebanese men', *Prostate Cancer,* pp.1-6.

Ellinghaus, D, Degenhardt, F, Bujanda, L, *et al.,* 2020, 'The ABO blood group locus and a chromosome 3 gene cluster associate with SARS-CoV-2 respiratory failure in an Italian-Spanish genome-wide association analysis', *MedRxiv*, (doi: https://doi.org/10.1101/2020.05.31.20114991).

Esteve, A, Permanyer, I, Boertien, D, Vaupel, JW 2020, 'National age and co-residence patterns shape COVID-19 vulnerability', *MedRxiv,* (https://doi.org/10.1101/2020.05.13.20100289).

Gard, PR 2010, 'Implications of the angiotensin-converting enzyme gene insertion/deletion polymorphism in health and disease: a snapshot review', *International Journal of Molecular Epidemiological Gene*tics, vol.1, no.2, pp.145-157.

Gattinoni, L, Coppola, S, Cressoni, M *et al.,* 2020, 'Covid-19 Does Not Lead to a "Typical" Acute Respiratory Distress Syndrome', *American Journal of Respiratory Critical Care Medicine*, vol.201, no.10, pp.1299-1300.

Geard, N, Glass, K, McCawa, JM, McBryde, ES *et al.,* 2015, 'The effects of demographic change on disease transmission and vaccine impact in a household structured population', *Epidemics*, vol.13, pp.56–64.

Ghosh, K, Sarkar, P, Chatterjee, D & Bandyopadhyay, AR 2018, 'Association of fat patterning, hypertension, and ACE I/D gene polymorphism: a study on two Tibeto-Burman linguistic groups of Tripura, North-east India', *MOJ Anatomy & Physiology*, vol.5, no.6, pp.368-371.

Gorbalenya, AE, Baker, SC, Baric, RS, *et al.,* 2020, 'The species Severe Acute Respiratory Syndrome-related Coronavirus: classifying 2019-nCoV and naming it SARS-CoV-2,' *Nature Microbiology*, vol.5, pp.536-544.

Hope-Simpson, RE 1970, 'First outbreak of Hong Kong influenza in a general practice population in Great Britain', *British Medical Journal,* vol.3, no.5714, pp. 74–77.

Huang, C, Wang, Y, Li X, *et al.,* 2020, 'Clinical features of patients infected with 2019 novel coronavirus in Wuhan, China', *Lancet*, vol.395, pp.497-506.

Jin, JM, Bai, P, He W *et al.,* 2020, 'Gender differences in patients with COVID-19: Focus on severity and mortality', *Frontiers in Public Health*, vol.8, no.152, pp.1-6.

John, AM 1990, 'Endemic disease in host populations with fully specified demography,' *Theoretical Population Biology*, vol.37, no.3, pp.455-471.

Kenyon, C 2020, 'ACE-1 I/D polymorphism associated with COVID-19 incidence and mortality: an ecological study,' *Preprints*, 2020040262 (doi: 10.20944/preprints202004.0262.v1).

Li, R, Richmond, P, Roehner, BM 2018, 'Effect of population density on epidemics', *Physica A: Statistical Mechanics and Its Applications*, vol.510, pp.713-724.

Lithander, FE, Neumann, S, Tenison E, *et al.,* 2020, COVID-19 in older people: a rapid clinical review. *Age and Ageing*. Vol.49, no.4, pp.501-515.

Livi-Bacci, M 1997, *A Concise History of World Population*, 2nd ed. Blackwell Publishers Inc., Oxford, UK.

Manfredi, P, Williams, JR 2004, 'Realistic population dynamics in epidemiological models: the impact of population decline on the dynamics of childhood infectious diseases. Measles in Italy as an example', *Math Biosciences*, vol.192, no.2, pp.153-175.

Nanda, A, Vura, NVRK, Gravenstein, S 2020, COVID-19 in older adults. *Aging Clinical Experimental Research,* pp.1-4.

Peeri, NC, Shrestha, N, Rahman, MS, *et al.,* 2020, 'The SARS, MERS and novel coronavirus (COVID-19) epidemics, the newest and biggest global health threats: what lessons have we learned*?' International Journal of Epidemiology,* (DOI: 10.1093/ije/dyaa033).

Pinheiro, DS, Santos, RS, VeigaJardim, PCB, *et al.,* 2019, 'The combination of ACE I/D and ACE2 G8790A polymorphisms reveal susceptibility to hypertension: A genetic association study in Brazilian patients', *Public Library of Science (PLoS), One* 14(8):e02212248.

Population density of India: Census 2011, Httpa://censusindia.gov.in. (21 July 2020).

Public Health England 2020, Disparities in the risk and outcomes of COVID-19. Public Health England; London.

Registrar General & Census Commissioner of India. Census of 2011.

Rocklov, J, Sjodin, H 2020, 'High population densities catalyze the spread of COVID-19', *Journal of Travel Medicine*, vol.27, no.3, (DOI: 10.1093/jtm/taaa038).

Saab, YB, Gard, P & Overall, A 2007, 'The geographic distribution of the ACEII genotype: a novel finding', *Genetic Research,* 89:259-267: (DOI https://doi.org/10.1017/S0016672307009019).

Shrestha, LB 2000, 'Population aging in developing countries', *Health Affairs*, vol.19, no.3, pp.204-212.

Singh, R, Adhikari, R 2020, 'Age-structured impact of social distancing on the COVID-19 epidemic in India', Preprint: https://www.researchgate.net/publication/340209224. (26 March 2020).

Song, P, Wei, Li W, Xie J, *et al.*, 2020, 'Cytokine storm induced by SARS-CoV-2', *Clinical Chimica Acta,* (*International Journal of Clinical Chemistry and Diagnostic Laboratory Medicine*), vol.509, pp.280–287.

Tikellis, C, Thomas, MC 2012, 'Angiotensin-Converting Enzyme 2 (ACE2) is a key modulator of the Renin-Angiotensin System in health and disease', *International Journal of Peptide Research and Therapeutics*, pp.1-8.

Verhagen, MD, Brazel, DM, Dowd, JB, Kashnitsky, I & Mills, MC 2020, 'Predicting peak hospital demand: demographics, spatial variation, and the risk of "hospital deserts" during COVID-19 in England and Wales', *Preprints,* (DOI: 10.31219/osf.io/g8s96).

Verity, R, Okell, LC, Dorigatti, I *et al.,* 2020, 'Estimates of the severity of coronavirus disease 2019: a model-based analysis', *The Lancet Infectious Diseases,* (DOI:https://doi.org/10.1016/S1473-3099(20)30243-7).

WHO 2019, Coronavirus disease, http://www.who.int/emergencies/diseases/novel - coronavirus-2019. (Accessed July 20, 2020).

WHO Coronavirus disease (COVID-19) advice for the public. https://www.Who.Int/emergencies/diseases/novel-coronavirus-2019/advice-for-public, (Accessed July 20, 2020).

World Population Ageing 2019. United Nations: http://creativecommons.org/licenses/by/3.0/igo/ (2019).

Yang, J, Zheng, Y, Gou, X, *et al.,* 2020, 'Prevalence of comorbidities and its effects in coronavirus disease 2019 patients: A systematic review and meta-analysis', *Internal Journal of Infectious Disease,* vol.94, pp.91-95.

Zaki, AM, Van, BS, Bestebroer, TM, Osterhaus, AD & Fouchier, RA 2012, 'Isolation of a novel coronavirus from a man with pneumonia in Saudi Arabia', *New England Journal of Medicine*, vol.367, no.19, pp.1814-1820.

Zhao, J, Yang, Y, Huang, H *et al.,* 2020, 'Relationship between the ABO blood group and the COVID-19 susceptibility', *MedRxiv*,(DOI: https://doi.org/10.1101/2020.
03.11.20031096).

Zhao, Y, Zhao, Z, Wang, Y, Zhou, Y, Ma, Y, and Zuo, W 2020, 'Single-cell RN expression profiling of ACE2, the putative receptor of Wuhan 2019- CoV', *BioRxiv,* 13 (DOI: https://doi.org/10.1101/2020. 01.26.919985).

Index

T

U

W

Y